Sustaining Design and Production Resources

John F. Schank

Jessie Riposo

John Birkler

James Chiesa

Prepared for the
United Kingdom's Ministry of Defence

 EUROPE

The research described in this report was prepared for the United Kingdom's Ministry of Defence. The research was conducted jointly in RAND Europe and the RAND National Security Research Division.

Library of Congress Cataloging-in-Publication Data

The United Kingdom's nuclear submarine industrial base.
    p. cm.
    "MG-326/1."
    Includes bibliographical references.
    ISBN 0-8330-3797-8 (pbk. vol 1)
    1. Nuclear submarines—Great Britain—Design and construction. 2.
Shipbuilding industry—Great Britain. 3. Military-industrial complex—Great
Britain. 4. Defense industries—Great Britain. I. Schank, John F. (John Frederic),
1946– II. Raman, Raj. III. Title.

    V859.G7.U55 2005
    359.9'3834'0941—dc22

                            2005010735

The RAND Corporation is a nonprofit research organization providing objective analysis and effective solutions that address the challenges facing the public and private sectors around the world. RAND's publications do not necessarily reflect the opinions of its research clients and sponsors.
RAND® is a registered trademark.

Cover design by Peter Soriano

Photo courtesy of Attack Submarine IPT, British Ministry of Defence

© Copyright 2005 RAND Corporation

Published 2005 by the RAND Corporation
1776 Main Street, P.O. Box 2138, Santa Monica, CA 90407-2138
1200 South Hayes Street, Arlington, VA 22202-5050
201 North Craig Street, Suite 202, Pittsburgh, PA 15213-1516
RAND URL: http://www.rand.org/
To order RAND documents or to obtain additional information, contact
Distribution Services: Telephone: (310) 451-7002;
Fax: (310) 451-6915; Email: order@rand.org

Preface

The design, engineering, and production of any complex system require special skills, tools, and experience. This is especially true for the industrial base that supports the design and construction of nuclear submarines. A single shipyard, Barrow-in-Furness, designs and builds the United Kingdom's nuclear submarines, and many of the vendors that support submarine construction, especially those associated with the nuclear steam-raising plant, are sole-source providers. The complexity and uniqueness of a nuclear submarine require special skills, facilities, and oversight not supported by other shipbuilding programmes.

Several recent trends have warranted concern about the future vitality of the United Kingdom's submarine industrial base. Force structure reductions and budget constraints have led to long intervals between design efforts for new classes and low production rates. Demands for new submarines have not considered industrial base efficiencies resulting in periods of feast or famine for the organisations that support submarine construction. Government policies have resulted in a reduction in the submarine design and management resources within the Ministry of Defence (MOD) in an effort to reduce costs. Yet the aforementioned production inefficiencies and increased nuclear oversight have resulted in increased costs.

Concerned about the future health of the submarine industrial base, the MOD asked RAND Europe to examine the following four issues:

- What actions should be taken to maintain nuclear submarine design capabilities?
- How should nuclear submarine production be scheduled for efficient use of the industrial base?
- What MOD capabilities are required to effectively manage and support nuclear submarine programmes?
- Where should nuclear fuelling occur to minimise cost and schedule risks?

This report addresses the first two issues.[1] The following companion reports address the last two issues:

- *The United Kingdom's Nuclear Submarine Industrial Base, Volume 2: MOD Roles and Required Technical Resources*, MG-326/2-MOD (forthcoming)
- *The United Kingdom's Nuclear Submarine Industrial Base, Volume 3: Options for Initial Fuelling*, MG-326/3-MOD.

This report should be of special interest not only to the Defence Procurement Agency and to other parts of the MOD but also to service and defence agency managers and policymakers involved in weapon system acquisition on both sides of the Atlantic. It should also be of interest to shipbuilding industry executives within the United Kingdom. This research was undertaken for the MOD's Attack Submarine Integrated Project Team jointly by RAND Europe and the International Security and Defense Policy Center of the RAND National Security Research Division, which conducts research for the US Department of Defense, allied foreign governments, the intelligence community, and foundations.

For more information on RAND Europe, contact the president, Martin van der Mandele. He can be reached by email at mandele@rand.org; by phone at +31 71 524 5151; or by mail at RAND Europe, Newtonweg 1, 2333 CP Leiden, The Netherlands.

---

[1] Some information specific to business-sensitive data is not cited herein but is made available in a restricted distribution version of this report.

For more information on the International Security and Defense Policy Center, contact the director, Jim Dobbins. He can be reached by email at James_Dobbins@rand.org; by phone at (310) 393-0411, extension 5134; or by mail at The RAND Corporation, 1200 South Hayes St., Arlington, VA 22202-5050 USA. More information about RAND is available at www.rand.org.

Contents

Figures

Tables

Summary

Since the end of the Cold War, the United Kingdom's defence budgets and military force structures have gotten much smaller. As a result, the defence industrial base has contracted as well. This industrial base must now be carefully managed to ensure that the capabilities required to support the nation's forces do not deteriorate to the point at which they cannot support defence requirements. An important factor in ensuring the sustainability of the industrial base is the scheduling of major weapon system acquisition programmes. Gaps in design and production can lead to the departure of experienced personnel to other industries and to the erosion of defence system production skills. This is particularly true of the nuclear submarine production base, for which special skills are required.

Given these concerns, the Attack Submarines Integrated Project Team within the Ministry of Defence (MOD) asked the RAND Corporation to examine the following questions pertaining to submarine design and production:

- What level of resources is needed to sustain a submarine design capability? When might, or should, the next design effort be undertaken? What actions should be taken to maintain submarine design capabilities during gaps between design efforts?
- How should submarine production be scheduled for efficient use of the industrial base? What are the implications of decisions regarding fleet size and production rate? How viable is the nonnuclear vendor base supporting submarine production?

Clearly, these questions can only be meaningfully addressed if a long view is taken. Definitive answers are thus not yet possible because aspects of the long-term future submarine fleet structure are unsettled. What we seek to accomplish here is to make some assumptions regarding that structure and work out the implications of acquisition options for the industrial base. In doing so, we develop an analytic framework that the MOD can apply again once more specifics are available.

The Fleet: Current and Planned

The UK submarine fleet now consists of 11 nuclear-powered attack submarines (SSNs) of the Swiftsure and Trafalgar classes and four nuclear-powered fleet ballistic missile submarines (SSBNs) of the Vanguard class. The Swiftsure boats are being retired over the next six years, and retirement of the Trafalgar class will begin shortly thereafter. Meanwhile, construction of the new Astute class of SSNs is under way. The first three boats of that class are under contract, and it has been announced that up to five more may be built. The first of class is now scheduled for delivery in 2009, with the next two boats following at 18-month intervals.

Sustaining the Design Base

The submarine design base is rapidly eroding. Demand for the design and engineering resources of BAE Systems Submarine Division at Barrow-in-Furness, which is designing and building the Astute class, is declining as the design of the first of class nears completion. Some professionals will be retained through the remainder of Astute-class production to provide design support to construction, but the number required will be fewer than that needed to sustain a viable nuclear submarine design base.

To sustain the United Kingdom's nuclear submarine design expertise, some minimum core of professionals must continuously work in that area. The number required varies with the domain of expertise: A few people may be enough to sustain submarine-specific expertise in some specialties, whereas some two-dozen persons may be needed to do so in disciplines such as marine engineering and systems engineering (see Figure S.1). The total number required across all domains is approximately 200. Even if the current distribution of skills among the BAE Systems submarine design force reflected that required to sustain the design base, the workforce could drop below this critical level in the near future without a new design programme.

Although there are other various options (which we will discuss below) for sustaining the 200-person submarine design core, the ideal way would be to soon commence the design effort for a new class of submarines. At this point, it is unclear when or even whether the United Kingdom will build another class of nuclear submarine. No decisions have been made regarding any programmes beyond the

Figure S.1
Number of People with Various Skills to Support a Nuclear Submarine Design Core

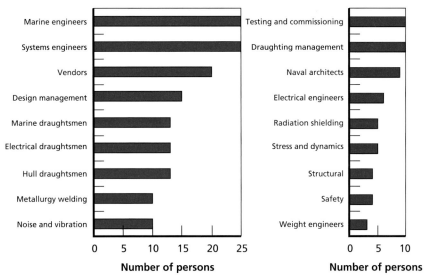

Astute class. However, the current Vanguard SSBN class could begin retiring as early as 2018. If the United Kingdom wishes to retain a submarine-based strategic nuclear deterrent, design of a follow-on SSBN class would have to start approximately 15 years prior to the desired in-service date for the replacement submarines. With retirements of the Vanguard class starting around 2018, the design for a follow-on class would have to begin immediately. The Astute-class boats will also need eventually to retire, and if the replacement for that class, now termed the Maritime Future Underwater Capability (MUFC), is to be a nuclear submarine, design would have to begin some 10 years in advance of delivery of the first of class.

For the purposes of developing and exercising a framework to permit the analysis of long-term programme planning and its implications for the industrial base, we make two assumptions: that the next submarine programme will be for a follow-on SSBN class and that the MUFC will be a class of nuclear attack submarines with the first of class delivered approximately 25 years after the first Astute-class boat becomes operational.[1] We assume design and production efforts for the MUFC will be similar to those for the Astute class, and we scale up the SSBN effort from that for the Astute class. (Should the next class be an SSN rather than an SSBN, some of the quantitative specifics presented here would differ somewhat, but our qualitative conclusions would not—and, of course, the analytic framework would remain valid.)

The first boat of the Vanguard class is now completing its mid-life refuelling. Originally, the Vanguard-class submarines were to have a life of 25 years, and that plan has not yet officially been changed, but the new reactor cores should permit operation until age 40. Were the Vanguard class to be retired at age 25, the design effort for the next of class would have to start immediately. That would reverse the near-term erosion of the design workforce. However, it would leave a gap of some six years between the major design efforts for the follow-on SSBN and the MUFC during which these pro-

---

[1] We also examine the implications if there is to be no follow-on SSBN class of submarines and if the MUFC programme can be moved forward.

gramme demands would be insufficient to support a core of expertise (see Figure S.2; Astute is omitted). Were the Vanguard class to be retired at age 40, that would close the gap between the SSBN design effort and that for the MUFC, but it would open an even larger gap in the near term. From a design base standpoint, the most desirable retirement age for the Vanguard class would be 30 to 35 years. That would largely close both the near-term gap between the Astute and SSBN classes and the far-future gap following the SSBN design effort (see Figure S.3; Astute is omitted).

Even if the Vanguard class is retired at 30 to 35 years of age, there may still be a period of time when the design core is inadequate in at least some of the specialties required to sustain expertise. And a retirement date that is not optimal for sustaining the design base may have to be chosen for some other reason. How might the design core be sustained through periods of slack demand? There are several possibilities:

- spiral development of the Astute class, that is, evolution of the Astute design as more boats are built to take advantage of new technologies and respond to changes in the threat the class is to meet
- continuous work on conceptual designs for new submarine classes, whether or not those classes are ever built
- design of unmanned undersea vehicles.

These options are not mutually exclusive; they could be exercised simultaneously. However, taken together, they could not by themselves adequately sustain a submarine design core. The work might not be quite enough, nor would it be entirely of the required character.

Collaboration with the United States or another submarine-producing country should also be considered. The United States confronts some of the same challenges in sustaining nuclear submarine design resources as does the United Kingdom. Design work on each country's submarine programmes could help sustain the other's

Figure S.2
Future Submarine Design Demands, Assuming Earliest Possible Start for
Design of New SSBN

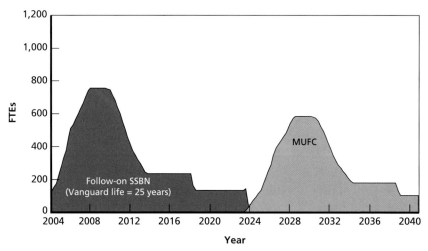

NOTE: Remaining Astute-related design work omitted to protect business-sensitive data.
RAND *MG326/1-S.2*

Figure S.3
Future Submarine Design Demands, Assuming Start for Design of New SSBN
to Replace Vanguard Class at Age 30

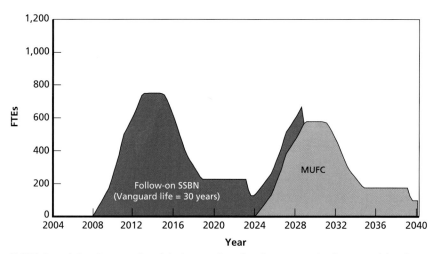

NOTE: Remaining Astute-related design work omitted to protect business-sensitive data.
RAND *MG326/1-S.3*

design core. If the United Kingdom could count on US assistance with submarine design, elements of the design core could be eliminated and the costs of sustaining it reduced (with some concomitant loss of independent design capability). Collaboration could also increase the interoperability of the countries' submarine forces. For collaboration to aid in sustaining the design core, however, the two countries cannot be at parallel positions in their submarine design programmes. At present, they are. The United States does not have a submarine design programme getting under way in the near future.

The MOD should also consider promoting collaboration between the UK organisations designing new submarine programmes and those supporting in-service submarines. There is not enough demand across nuclear submarine design, production, and support to continuously sustain large numbers of design professionals at each organisation. Support organisation designers and engineers could be part of any new submarine design programme, bringing their general knowledge of submarine design plus their specific knowledge of the support of in-service submarines. Likewise, designers and engineers from the Barrow shipyard could aid in the in-service support of the Astute class.

Sustaining and Maximising the Efficiency of the Production Base

As mentioned above, the United Kingdom's submarine production base will be sustained for the next several years by the current Astute contract. That leaves two questions: What happens after that, and could the Astute boats be built more slowly than the prevailing 18-month production 'drumbeat'?[2]

The answer to both questions again depends on when construction of the next submarine class begins. If the next class is to be a follow-on SSBN class, that would in turn depend on when the Van-

---

[2] We use the term *drumbeat* throughout this report to represent a consistent production rate. An 18-month drumbeat suggests the construction of a new submarine begins every 18 months.

guard class retires. The conclusions here reflect those of the design base analysis. Production consistent with a 25-year retirement date leaves too great a gap after SSBN construction concludes and before MUFC construction starts. It also overlaps Astute production too much, giving rise to a peak in the demand for submarine construction resources that the Barrow shipyard would have trouble satisfying. A 40-year retirement date implies simultaneous production of the next SSBN and MUFC classes following a long production gap after Astute construction ends. A 30- to 35-year retirement age for the Vanguard class provides the opportunity for continuous submarine production into the distant future.

In the event of a 30-year SSBN replacement schedule, the end of follow-on SSBN construction could be timed for a smooth overlap with the start of MUFC construction. The nature of the overlap at the start of SSBN construction depends on how fast the SSBNs and Astute-class boats are built. Slowing down Astute production and building the SSBNs relatively quickly would result in a reduced demand for production resources between 2010 and 2020, followed by a ramp-up to meet the SSBN demand. An almost even demand profile could be achieved with a relatively fast Astute drumbeat of 18 months, followed by a slow SSBN drumbeat of 36 months (see Figure S.4). The transition from Astute production to SSBN production would then occur between 2015 and 2020. It is noteworthy that the overlaps allowed by long-term production planning smooth not only the total production demand but also the demand for broad skill categories such as hull construction and outfitting.

Starting new submarine programmes after gaps in submarine production at Barrow will incur substantial costs and risks. If there is no follow-on to the Vanguard class, production of the remaining Astute-class boats may have to be stretched (i.e., built at a slower drumbeat) and the start of the MUFC programme accelerated. Even with these actions, maintaining a force size of eight SSNs could be prohibitively expensive. Unit production costs could be reduced if more SSNs are built either by increasing the fleet size or by retiring active submarines early. However, both of these strategies could lead to higher total nuclear submarine production and through-life costs.

Figure S.4
Workforce Demands at Barrow (18-month SSN/36-month SSBN Drumbeats)

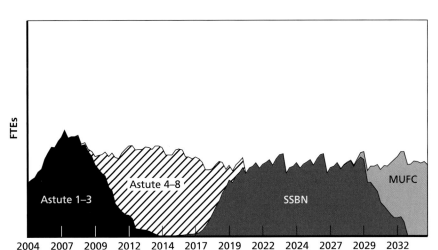

RAND *MG326/1-S.4*

Smoothing out the demand on submarine production resources would allow the industrial base to operate at peak efficiency and could reduce production costs by 5 to 10 percent per boat. However, demand-smoothing is not likely to be the only factor considered in scheduling production. It may be necessary for other reasons to build the last five Astute boats more slowly or even to delay the start of their construction. If so, the resulting valleys in production demand at Barrow could be filled with work on other ship programmes. In the coming years, the United Kingdom will be building the Type 45 surface combatants, the Future Aircraft Carrier (CVF), and the Maritime Reach and Sustainability (MARS) ships. These simultaneous programmes will tax the nation's shipbuilding capacity, and some contribution from the Barrow yard to the effort may be helpful.

It should be kept in mind that an 18-month production drumbeat is quick only in relative terms. It is still slow enough that we sought to determine whether any of the nonnuclear vendors to the

Barrow yard[3] might be having difficulty surviving and whether their loss could give rise to critical supply gaps. We found that, although some firms had indeed dropped out of the vendor pool, replacements had been found or alternatives were available. Furthermore, no companies still in the vendor base were both irreplaceable and in danger of failing. However, stretching production drumbeats beyond 24 months will require actions on the MOD's part to sustain portions of the vendor base.

Recommendations

From the preceding analysis, we infer a number of actions that the MOD could take to ensure that the United Kingdom's nuclear submarine industrial base is sustained and operates efficiently:

- Decide as soon as possible whether there will be a next-generation SSBN class and when it will be designed and built. This decision is needed to inform any further actions to sustain the design base and schedule remaining Astute production to maximise efficiency. If the Vanguard class is not to be replaced, then planning should begin for an early follow-on SSN class if the submarine design base is to be sustained.
- Plan on annual investments to sustain a core of submarine-specific design resources. A core of 200 designers, engineers, and draughtsmen would require annual funding of perhaps £15 million. That would permit the core to participate in meaningful work such as spiral development of Astute and continuous conceptual development when there is no new submarine design effort.
- Take steps towards collaboration. Although there is not a near-term US submarine programme to support the UK design core during a lapse in demand, collaboration might be beneficial in

---

[3] Nuclear vendors are being considered in a separate analysis conducted for the MOD by another organisation.

the future. It would thus be prudent to begin talks towards a small collaborative effort, with respect to either spiral development of the Astute and US Virginia classes or the design of follow-on SSBN classes in both countries. Meanwhile, the MOD should encourage collaboration between BAE Systems' Barrow-in-Furness shipyard and the contractors employing engineers and draughtsmen for in-service support.

- Decide on the timing of construction for the next Astute-class contract. To sustain and make most efficient use of the submarine production base, an 18-month (or, at most, 24-month) drumbeat should be employed, with no additional break between the third and fourth boats of the class (i.e., the last boat under the current contract and the first under the next). If slower production or a delay is required for some reason, the MOD should allocate some of the work from such programmes as the CVF or MARS to the Barrow yard to level the load there.
- Take action to support nonnuclear vendors. Although there are typically ongoing challenges in maintaining the vendor base that supports shipbuilding programmes (because some suppliers decide to leave the industry or to forego naval contracts), no vendor problems are currently foreseen that cannot be solved. However, the MOD might preempt future problems by placing orders for multiple ship sets of equipment, encouraging other shipbuilding programmes to use the submarine programme vendors when possible, and collaborating with the United States to identify common vendors.

Acknowledgements

This research could not have been accomplished without the assistance of many individuals. Muir Macdonald, leader of the Attack Submarine Integrated Project Team (IPT), supported and encouraged the work. Numerous individuals in the Attack Submarine IPT offered information, advice, and assistance. Nick Hunt and Stephen Ranyard were especially helpful, providing background on the Astute programme and offering constructive criticism of interim findings and documentation. If we were to single out one individual who supported us in extraordinary ways, it would be Helen Wheatley, who provided data and information and facilitated our interactions with multiple organisations.

Commodore Paul Lambert and Commodore Mark Anderson, former and current directors, respectively, of Equipment Capability, Underwater Effects, offered advice and guidance on nuclear submarine requirements. Commander Nigel Scott of their Directorate was especially helpful during the course of the research. Commander JJ Taylor of the Submarine Support IPT provided data on the planned retirements and maintenance actions for the current fleet of Royal Navy submarines. Peter Duppa-Miller, Director of the Submarine Library, shared his extensive historical knowledge of UK submarine programmes and provided many documents that helped in the research. Colin Bennet of the Pricing and Forecasting Group offered valuable suggestions that strengthened the report.

Many individuals at BAE Systems Submarine Division shared their time and expertise with us and provided much of the data that were necessary to perform the analyses. Notable were Murray Easton, managing director at Barrow; Huw James; Duncan Scott; Mark Dixon; and Michael Wear. We are deeply thankful for their assistance. Steve Ludlam, director of submarines at Rolls-Royce Naval Marine, and Ken Grove, managing director of Strachan & Henshaw, provided valuable insights into the problems faced by the vendors that support the nuclear submarine programmes. Steve Ruzzo, Hank Rianhard, and Tod Schaefer of the Electric Boat Corporation and Becky Stewart, Charlie Butler, and Chris Vitarelli of Northrop Grumman Newport News provided data and insights on US submarine programmes.

At RAND, Giles Smith offered valuable insights and suggestions that greatly improved the presentation of the research; Deborah Peetz provided her typically excellent support to the overall research; and Phillip Wirtz edited the final document.

Special thanks go to USN RADM (ret) Malcolm MacKinnon, who more than a decade ago introduced us to the intricacies of submarine design and construction. Many of his thoughts and words are reflected in the analysis of sustaining submarine design resources.

The abovementioned individuals helped us with functional information and suggested some implications. We, however, are solely responsible for the interpretation of the information and the judgements and conclusions drawn. And, of course, we alone are responsible for any errors.

Abbreviations

BES	Babcock Engineering Services
CVF	Future Aircraft Carrier
DML	Devonport Management Limited
FTE	full-time equivalent
IPPD	integrated product and process design
IPT	Integrated Project Team
LPD	landing platform, dock
MARS	Maritime Reach and Sustainability
MOD	Ministry of Defence
MUFC	Maritime Underwater Future Capability
PAD	project acceptance date
PLC	Public Limited Company
SQEP	suitably qualified and experienced personnel
SSBN	nuclear-powered fleet ballistic missile submarine (ship submersible ballistic nuclear)
SSN	nuclear-powered attack submarine (ship submersible nuclear)
UUV	unmanned undersea vehicle
VSEL	Vickers Shipbuilding and Engineering Limited

CHAPTER ONE
Introduction

During the Cold War, with a clearly defined enemy and clearly defined threats, the major allied nations such as the United Kingdom and the United States maintained fairly large force structures and placed significant demands on their defence industrial base. However, since the end of the Cold War, defence budgets and force structures have become much smaller, causing the defence industrial base to contract and change as well. Now, nations like the United Kingdom must closely monitor and manage their industrial base to ensure that the capabilities required to support their forces do not deteriorate but will be available when needed. Activities that might be modified to sustain a robust industrial base might include the scheduling and assigning of new design and construction programmes.

Industrial base challenges are especially difficult for the design and production of nuclear submarines. In the United Kingdom, submarines are the only types of ships that use nuclear propulsion; partly as a result, the personnel skills and disciplines necessary for nuclear submarine design and production are unique in the ship-building industrial base. The recently publicised cost and schedule problems with the Astute programme are a manifestation of the difficulties that can arise in the nuclear submarine industrial base. The gap between the end of the preceding submarine programme—Vanguard—and the start of Astute has been postulated as one of Astute's problems. Such gaps in design and production present problems, since many of the necessary disciplines and skills cannot be

1

maintained by other, non-submarine programmes. (Other possible problem sources include a lack of sufficient design resources and frequent management changes at the shipbuilder.[1])

The UK submarine industrial base is facing potential future gaps in submarine design and production programmes. If there is no successor to the Vanguard class, there may be a 20-year gap between submarine design efforts—that is, between the end of the design of the Astute class and the start of the design of the follow-on Maritime Underwater Future Capability (MUFC) class (which may not even be a submarine). Also, given the small numbers of attack submarines in the Royal Navy inventory and the possibility of no successor to the Vanguard class, there are likely to be gaps of several years between the end of the Astute production and the start of production for the next class.

Questions, therefore, arise concerning how best to maintain submarine design and production capabilities in this era of declining defence budgets and force structures.[2] Answers to these questions must consider both the Ministry of Defence (MOD) and all the organisations that comprise the nuclear submarine industrial base, including prime contractors, shipbuilders, component and equipment vendors, and the organisations that support in-service submarines.

---

[1] There could be valuable lessons learned from the design and production gaps between the Vanguard class and the start of the Astute class. Unfortunately, since the design is not finished and construction of the first of class will not be complete until 2008, it is too early to understand the full implications of the gap in submarine work. Also, there are numerous interacting factors that contribute to the cost and schedule problems faced by Astute. It would be difficult to filter out the impact of the gap from the influence of these other factors.

[2] The United States has faced similar questions over the last decade. See, for example, John Birkler, John Schank, Giles K. Smith, Fred Timson, James Chiesa, Marc D. Goldberg, Michael Mattock, and Malcolm MacKinnon, *The U.S. Submarine Production Base: An Analysis of Cost, Schedule, and Risk for Selected Force Structures*, Santa Monica, Calif., USA: RAND Corporation, MR-456-OSD, 1994; John Birkler, Michael Mattock, John Schank, Fred Timson, James Chiesa, Bruce Woodyard, Malcolm MacKinnon, and Denis Rushworth, *The U.S. Aircraft Carrier Industrial Base: Force Structure, Cost, Schedule, and Technology Issues for CVN 77*, Santa Monica, Calif., USA: RAND Corporation, MR-948-NAVY/OSD, 1998; and John Schank, John Birkler, Eiichi Kamiya, Edward Keating, Michael Mattock, Malcolm MacKinnon, and Denis Rushworth, *CVX Propulsion System Decision: Industrial Base Implications of Nuclear and Non-Nuclear Options*, Santa Monica, Calif., USA: RAND Corporation, DB-272-NAVY, 1999.

Study Objectives and Research Approach

Given the concerns surrounding the vitality of the submarine industrial base, the Attack Submarine Integrated Project Team (IPT) within the MOD asked RAND Europe to examine the following questions pertaining to submarine design and production:

- What level of resources is needed to sustain a submarine design capability? When might, or should, the next design effort be undertaken? What actions should be taken to maintain submarine design capabilities during gaps between design efforts? (See Chapter Two.)
- How should submarine production be scheduled for efficient use of the industrial base? What are the implications of decisions regarding fleet size and production rate? How viable is the non-nuclear vendor base supporting submarine production? (See Chapter Three.)

Clearly, these questions can only be meaningfully addressed if a long view is taken. Definitive answers are thus not yet possible: important aspects of the long-term future submarine fleet structure are unsettled. What we seek to accomplish here is to make some assumptions regarding that structure and work out the implications of acquisition options for the industrial base. In doing so, we develop an analytic framework that the MOD can apply again once more specifics are available.

Because of the uncertainty surrounding future UK submarine programmes and force structures, our analysis considered various options, including

- a follow-on to the Vanguard class as the next new submarine programme
- no new submarine programme until MUFC (i.e., no follow-on to the Vanguard class)
- extending the operational life of the Vanguard class

- various numbers of nuclear-powered attack submarines (SSNs) and nuclear-powered fleet ballistic missile submarines (SSBNs) in the Royal Navy force structure
- various times between production starts of new submarines
- different start dates for the next submarine after the first three Astute boats.

In examining the issues listed above, we gathered information from a variety of sources, which included a wide range of interviews. In addition to interacting with several organisations within the MOD, we talked extensively with various organisations that are part of the nuclear submarine industrial base, including the BAE Systems Submarine Division in Barrow-in-Furness, Devonport Management Limited (DML), Rolls-Royce Naval Marine, and Strachan & Henshaw. We gathered insights and information from US nuclear submarine organisations, including the Program Executive Officer for Submarines as well as the Electric Boat Corporation and Northrop Grumman Newport News, the two organisations that design and build all US nuclear submarines. We conducted an extensive literature review of past studies addressing the US and UK nuclear submarine industrial base.[3] Finally, we performed quantitative analyses supported by databases and models constructed for previous analyses of US and UK shipbuilding studies.

Before we convey our findings regarding submarine design and production issues, it will be useful to set the scene with a description of the submarine industrial base in the United Kingdom.

The UK Submarine Industrial Base

The design and production of nuclear submarines in the United Kingdom is accomplished by BAE Systems and its Submarine division, which is located at the Barrow-in-Furness shipyard in northwest

---

[3] The review of reports addressing UK nuclear submarine issues was greatly facilitated by Peter Duppa-Miller, head of the UK submarine library.

England.[4] BAE Systems is the prime contractor for Astute design and production. It is also the design authority[5] for the class—i.e., it is primarily responsible for each boat's design and has the authority to determine whether the boat is being produced to the requirements of the design (with the exception of certain safety-related aspects, which must still be certified by the MOD).

Numerous vendors provide various equipment and materiel to Barrow to support submarine construction. Rolls-Royce Naval Marine is the sole provider for the nuclear steam-raising plant and acts as the delegated design authority for the Astute-class nuclear propulsion system. Two of the larger nonnuclear vendors, in terms of the cost of the equipment provided, are Strachan & Henshaw, which is responsible for the weapons handling and launch systems, and Thales Underwater Systems, which provides the sonar and other combat systems. There are more than 30 vendors that provide equipment or materiel worth in excess of £1 million each for the current Astute contract and hundreds of other vendors that provide smaller quantities of parts and materiel.

The Royal Navy fleet currently includes a declining number of Swiftsure-class SSNs,[6] seven Trafalgar-class SSNs, and four Vanguard-class SSBNs. The seven Trafalgar-class submarines have their homeports at Her Majesty's Naval Base Devonport in Plymouth, the largest naval installation in Western Europe. DML is responsible for the maintenance of all ships at Devonport; it is also the sole facility licensed to perform mid-life refuelling and end-of-life

---

[4] A historical description of submarine programmes in the United Kingdom and the role of the Barrow shipyard is provided in Appendix A.

[5] Design authority includes maintaining the top-level specifications for the submarine and ruling on emergent design, production quality, and other compliance issues that occur during construction. Ultimately, having design authority means ensuring that the product is complete and meets its requirements.

[6] Six Swiftsure-class submarines were produced between 1969 and 1979. Four currently remain in the inventory, and all will be retired within the next six years.

deactivations of nuclear submarines.[7] The Swiftsure- and Vanguard-class submarines are based at Faslane, part of Her Majesty's Naval Base Clyde in Scotland. The Astute class will also be based at Faslane. Babcock Engineering Services (BES) manages and performs the submarine maintenance functions at Faslane. BES is not licensed to perform nuclear refuellings or deactivations but can do the other scheduled maintenance actions for the Swiftsure- and Vanguard-class boats. DML and BES, along with the MOD's Submarine Support IPT, work together to develop maintenance plans and work packages for the support of in-service submarines.

---

[7] See companion document, *The United Kingdom's Nuclear Submarine Industrial Base, Volume 3: Options for Initial Fuelling*, MG-326/3-MOD, for a discussion of DML's facilities and operations.

CHAPTER TWO

Maintaining Nuclear Submarine Design Resources

The design and engineering[1] of any complex system requires special skills, tools, and experience. Of all military and commercial ships, a nuclear-powered submarine presents the greatest design challenge. The unique operating environment and characteristics of a nuclear submarine impose special demands on designers and engineers. These individuals need unique skills to address the ability to operate in three dimensions, the requirement to submerge and surface, the fine degree of system integration due to weight and volume limitations, and the use of nuclear propulsion. These skills are not found or maintained in the design of other UK military or commercial ships.

In this chapter, we examine various issues associated with maintaining the capability to design nuclear submarines. Specifically, we address the following questions:

- What resources are required to design a new submarine class? What is required to sustain a design capability?

---

[1] Typically, 'design' is the creative activity encompassing naval architecture and all aspects of marine engineering necessary to produce a new concept or design a major modification to an existing one. In contrast, 'engineering' is the application of engineering tools and principles to solve specific problems for the designer and to provide the support for the translation of the design to production. We group these two activities together in our discussions of the design process. For a review of the phases of the design process, see Appendix B.

- How should future design programmes be timed to sustain design resources?
- What are options for sustaining a minimum set of design resources between new design programmes?

First, however, we elaborate on the nuclear submarine design challenge: Why is it difficult to maintain nuclear submarine design resources?

Problems in Maintaining Resources

Any ship design programme requires a mix of skills, including those of naval architects, systems engineers, and marine engineers, plus designers and engineers skilled in specific systems such as electrical and mechanical systems, structure, and stress and dynamics. In addition, workers experienced in project management and test and commissioning are required, as well as a wide range of draughtsmen. The design of a nuclear submarine adds a set of unique skills to these general ship design and engineering resources (see Table 2.1). These skills include nuclear propulsion, noise and vibration (e.g., acoustics), and radiation shielding. It is these unique skills that can only be sustained through nuclear submarine design efforts.

Table 2.1
Skills Required for Nuclear Submarine Design

Naval architecture	Combat systems engineering
Mechanical engineering	Safety and operability engineering
Electrical engineering	Test and commissioning
Structural engineering	Design management
Noise and vibration	Applications engineering
Stress and dynamics	Draughting
Weight engineering	–electrical
Metallurgy and welding engineering	–marine
Radiation physics and shielding	–hull
Systems engineering	–management
Nuclear propulsion	

Maintaining submarine specific skills is not difficult when there is a sufficient overlap of new design efforts. Historically, as one design programme was winding down, the design for the next new submarine started. Figure 2.1 shows the time between the project acceptance dates (PADs)[2] for the first boat in various classes of nuclear submarines designed in the United Kingdom (the PAD for the Upholder class, a diesel submarine, is added for completeness). The time between the PAD for the first boat in one class and the first boat in the following class provides a measure of the overlap between design programmes.[3] Until the Astute class, the time between PADs for the first of class was eight to 10 years. This provided some overlap

Figure 2.1
Time Between First-of-Class Project Acceptance Dates

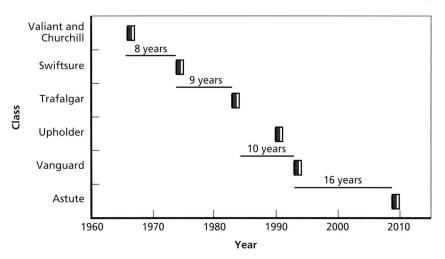

---

[2] Final project acceptance comes after sea trials and before commissioning.

[3] We use the PAD for the first of class as a proxy measure of design effort, since data were not available on the design periods for each class of submarines. Also, the government, not the shipbuilder, did much of the initial design work for submarines up through the Vanguard class. That trend changed for the Astute class. Although there is some level of design work associated with follow-on construction, we assume the vast majority of the design effort for a class is complete when the first of class is delivered.

between design programmes, since it takes, on average, 10 years to design and build the first of class for an SSN and 15 years for an SSBN. The overlap permitted designers and engineers to move from one programme to the next without the need to commit to other work.

The problem occurs when there is a gap between the end of one design programme and the beginning of the next. There has been a gap of 16 years between the PAD for the Vanguard and the PAD for the Astute. Such a gap between new design programmes makes it difficult to maintain the nuclear submarine design resources. When a gap occurs, skilled workers are transferred to other shipbuilding programmes, are placed on overhead, or leave the nuclear submarine design base altogether. Because they are highly trained and have skills that are in demand in other industries, nuclear submarine designers and engineers, especially electrical and mechanical engineers and draughtsmen, typically have little problem finding positions outside the shipbuilding industry. As these experts commit to other career paths, the loss to the national submarine design base becomes permanent. Such a net erosion of design skills associated with a programme gap is viewed as one of the causes of the problems experienced by the Astute programme.

Hiring new people to replace lost personnel is difficult. Prospective new hires into nuclear submarine design often view it as an industry with a limited future and large uncertainties for steady employment. Also, the Barrow area is somewhat remote with very few other employment opportunities for designers or engineers. This compounds the difficulty in attracting new hires and in rehiring those who obtain other employment during periods of decreased workloads.

Furthermore, experience is a fundamental requirement in all phases and at all levels, particularly with those key individuals who must lead and oversee the design effort. The combination of experience and leadership in total nuclear submarine design synthesis rests with a few special individuals—some in the private sector, some employed by the Royal Navy or the MOD. These leadership functions require 10 to 15 years to develop. Thus, this talent is critical to maintain and very difficult to recruit.

The Barrow shipyard is facing a watershed in sustaining nuclear submarine design resources. As shown in Figure 2.2, the demand for direct design and engineering resources has been falling. Demand will continue to erode through the end of the current Astute contract (which covers boats 1, 2, and 3). Further design should start immediately, or it will take several years and significant funding to reconstruct the nuclear submarine design base for the next new programme.

Resources Required for a New Design Effort

In identifying potential gap-induced shortfalls in the submarine design base, the first question to answer is: What is needed? That is,

Figure 2.2
Demand for Design Resources at Barrow

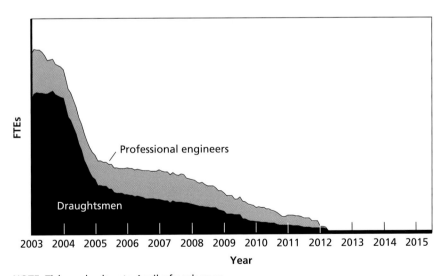

NOTE: Tick marks denote April of each year.
RAND MG326/1-2.2

how many engineers and draughtsmen are required? And how does the answer to that question change over the course of design?

In addressing these questions, we interviewed individuals involved with previous UK submarine design programmes as well as knowledgeable people at Barrow, Northrop Grumman Newport News, and Electric Boat. We supplemented the information gathered from these interviews with historical data on previous UK and US submarine design programmes.

To estimate the resources required, we had to settle on the general character of the next class of submarines to be designed. Given the projected retirement dates of the Vanguard and Astute classes, we initially assume the next class will be SSBNs. It should be kept in mind that no decision has been made as to whether there will be a new SSBN class to replace the current one.

Assuming there is an enduring requirement for nuclear attack submarines, a decision to not have a follow-on to the Vanguard class would mean the next new submarine design effort would be for an SSN. Currently, the MOD envisions a system providing what it terms an MUFC as the successor to the Astute class. With a projected 25-year life for the Astute class, the MUFC design effort would start in approximately 2024 to produce the first MUFC boat in approximately 2034. With no follow-on to Vanguard, there would thus be a gap of approximately 15 years between the PAD of the first Astute-class submarine and the start of the next submarine design. Such a large gap would make it very difficult and costly to reconstitute the nuclear submarine design base.

To preserve nuclear submarine design resources, a new SSN programme would have to start in the 2010 time frame and subsequent new SSN designs would have to start approximately every 10 years to avoid design gaps (see Figure 2.1 for a measure of historical gaps in design programmes). Longer intervals would result in the atrophy of design skills (although, as we will discuss, a core of nuclear submarine design resources might be sustained). Ten-year intervals are feasible, but they would lead to several small classes of SSNs in the fleet at any one time, most likely built at higher cost per boat than larger classes would be.

Although the analysis that follows is based on the assumption that the next new design effort is for an SSBN, the general findings and recommendations hold true if the next design effort is instead for a new SSN. The important issue is the timing, not the character, of the next new programme.

In the following analysis, we also assume, except where otherwise stated, an evolutionary design rather than a revolutionary one. We estimate that the design of a new evolutionary SSBN would span approximately 15 years from Initial Gate[4] to PAD, thus approximating the historical average. A revolutionary design would require an additional two to three years as well as an increase in the level of design resources required. The time from Initial Gate to PAD for an evolutionary SSN would be approximately 10 years.

Those times do not include the exploratory conceptual studies examining alternative design options that would precede Initial Gate. They begin with what is commonly referred to as preliminary design, in which subsystem configurations and alternatives are examined and analysed for military effectiveness, affordability, and producibility. The times encompass subsequent design phases, concluding with engineering support during construction.

Figure 2.3 shows our estimate of the number of professional engineers and draughtsmen required during the 15-year period for evolutionary SSBN design. The FTEs required peak at about 750 in the sixth year after Initial Gate. After about two years at that level, the demand drops until only 100 to 150 engineers are needed for support during construction. This estimate totals approximately 6,300 man-years (approximately 11 million man-hours) of design and engineer-

---

[4] Initial Gate is the decision point between the Concept stage and the Assessment stage of the Concept, Assessment, Demonstration, Manufacture, In-Service, and Disposal (CADMID) cycle. It is 'where the Assessment stage is approved and time, cost and performance boundaries of validity will be noted for the project as a whole'. UK Ministry of Defence, *The Smart Acquisition Handbook*, Edition 5, Director General Smart Acquisition Secretariat, January 2004.

Figure 2.3
Design Resources Needed for New SSBN Programme

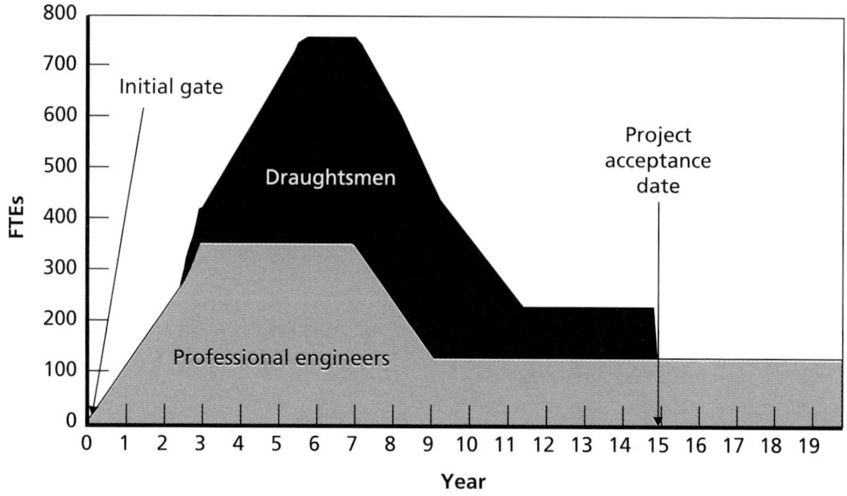

ing work,[5] which is 30 percent greater than the design effort for the Astute class (approximately 8.5 million man-hours for stages 1 through 3).

Various factors could lead to lower or higher resource levels needed for the design of a follow-on SSBN. As stated above, a revolutionary design that did not build on Astute-class technology would increase the design and engineering hours.[6] Enhancements to computer-aided design tools may decrease the hours required or change the mix of skills. Future design software may include a knowledge base of embedded rules. For example, Electric Boat estimates that its

---

[5] Figure 2.3 does not include the combat systems designers and engineers. We estimate that approximately 1.5 million man-hours are needed for the combat system design effort for the next new submarine class. This estimate is similar to the effort for the Astute class. We do not include combat system designers and engineers in our estimate of a core submarine design capability, since these skills are available in the general shipbuilding design base.

[6] The level of resources for US submarine design programmes are typically two to three times the level of resources required for UK programmes. Part of this difference is due to the revolutionary nature of US programmes as compared with the typically evolutionary nature of UK programmes.

next new design effort will require only 60 to 70 percent of the resources required for the Virginia programme because of advances in computer software and lessons learned from previous programmes.

It is important to note that Figure 2.3 shows the course of design *demand* over a programme's duration. Ideally, on the supply side, the submarine design force would not build up from zero. The origin and experience of the 750-member SSBN design team should instead be characterised as shown schematically in Figure 2.4. The team should be structured around a continuously sustained core of submarine-specific designers and engineers. A year or two before the Initial Gate of a new programme, personnel drawn from the general ship-design skill pool would begin to augment the core. This augmentation would continue until the peak demand for design resources occurs, approximately six years after Initial Gate (see Figure 2.3). Ideally, these professionals would be recovered from previous submarine design efforts. That is, when a submarine design programme ends and no new submarine design work is starting, the majority of the design team would be reassigned to other, non-submarine ship design programmes.[7] When a new submarine design programme starts, these designers, engineers, and draughtsmen would transition back to the new design team.[8] In actuality, persons with submarine experience might have moved on to other yards and might well not be easily recoverable. Given that, and given the time that might have elapsed since the last submarine programme, it is safer to assume that persons transferring in from other design programmes have only general skills. Depending on the number of people available with these general ship design skills, the submarine design team would recruit individuals from outside the ship design base. (These

---

[7] This assumes that there is other, non-submarine design work available for designers and engineers as a submarine design programme winds down. Thus, in addition to maintaining submarine specific design skills (the centre core of Figure 2.4), there is a need to sustain general design and engineering skills in UK shipbuilding (the middle ring of Figure 2.4).

[8] There are approximately 1,000 designers and engineers at DML and BES who provide support to in-service submarines. These DML and BES resources should be considered when forming a new submarine design team. They not only provide design skills but can also bring experience from supporting the in-service submarines to the design process.

people might have had ship design experience at some point, however; they are represented by the growth ring in Figure 2.4.)

The peak design force would be busy for a few years until construction commences on the first submarine in the class. The design team would then start to disperse, sending people back to the general ship design pool. Eventually, the design team would be composed solely of the core personnel.

The core is the critical part of the submarine design team. These designers, engineers, and draughtsmen should spend most of their time continuously on submarine design work. The special nature of some skills and the need for experienced personnel require this core as a base on which to build new teams.

We estimate the size of this continuous core of submarine-specific people at around 200 designers, engineers, and draughtsmen.

Figure 2.4
Composition of Nuclear Submarine
Design Team

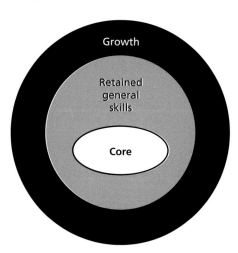

An estimate of the number of people in various skills is shown in Figure 2.5. In each skill area, there would be a mix of senior, mid-level, and junior people. Note that the list includes designers and engineers working for the major vendors, such as Rolls-Royce Naval Marine, Strachan & Henshaw, and Thales Underwater Systems.[9] As with submarine design in general, these providers of major systems also require a core of design resources to be maintained between new submarine design programmes.[10]

Figure 2.5
Number of People with Various Skills to Maintain a Nuclear Submarine Design Core

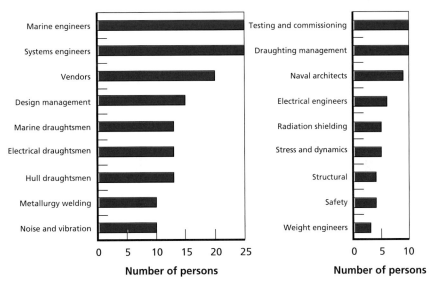

RAND MG326/1-2.5

---

[9] The design resources required by the major vendors are not included in our estimates for a new submarine programme shown in Figure 2.3.

[10] Designers and engineers from DML and BES should also be considered when forming the core.

Timing the Design Resource Demands

We can now lay out the programmatic resource demands such as that shown in Figure 2.3 on a timeline so we can see when demand might overwhelm supply or drop to zero, leaving a gap. Such projections, of course, are highly uncertain, but the results fall into patterns that we think are instructive, even if the specific numbers are best regarded as approximate.

Because there is some flexibility in retirement dates for the Vanguard class, we can use the temporal profile of demands to determine when the next future design effort *should* occur if sustaining the design base were a primary concern. However, we should not restrict ourselves to considering only the gap between the Astute and follow-on SSBN design, but also the gap between the latter and the design of the next attack submarine system. As stated above, design of the MUFC should start in approximately 2024.

The HMS *Vanguard* was brought into service with the Royal Navy in 1993. The boat is currently completing its mid-life refuelling and refit. Based on the initial plans for a 25-year operating life, a replacement for the Vanguard would be required in 2018. Assuming a 15-year design window from Initial Gate to PAD, the follow-on SSBN design should have started in approximately 2003. We now lay the following out on the timeline just given: the projected SSBN design profile from Figure 2.3 and a 77-percent-scale version of that profile for a hypothetical MUFC demand profile (see Figure 2.6).

Obviously, the required start date for a replacement at 25 years has already passed. However, the 25-year Vanguard operational life is problematic in ways that apply to immediate-future starts as well. First, given that it is unlikely that the submarine design base has sufficient resources to support two design programmes, having SSBN design under way now and through the near future leaves no window for further development of the Astute-class boats.

Second, starting the future SSBN class design so early means finishing it early, resulting in a gap until the beginning of the MUFC

Figure 2.6
Future Submarine Design Demands, Assuming Earliest Possible Start for Design of New SSBN

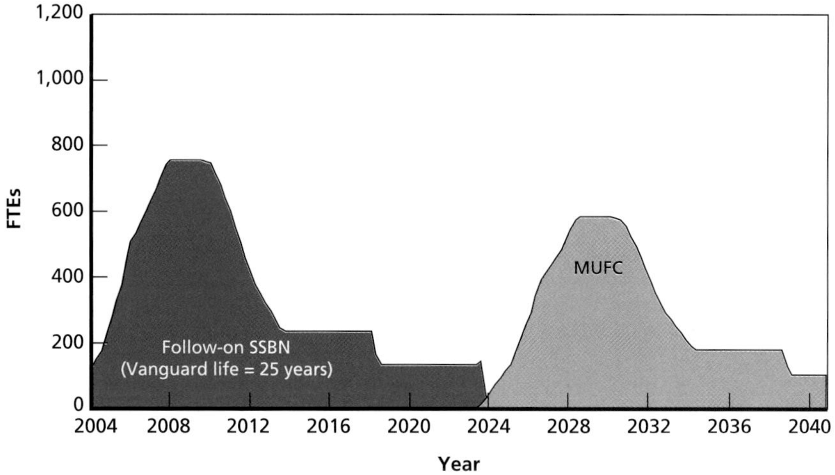

NOTE: Remaining Astute-related design work omitted to protect business-sensitive data.
RAND *MG326/1-2.6*

design. As shown in Figure 2.6, the demand for submarine design professionals falls below 200 in approximately 2018 and stays there for several years. It is also unclear whether the 230 engineers and draughtsmen needed in the years before 2017 will include the 'right' 200 for maintenance of core design specialties, so the gap may be measured as *at least* six years.

Finally, as shown in Figure 2.7,[11] if a new submarine design programme were under way in the immediate future, it would compete for overall design resources with several other programmes. The total design demand for those programmes (along with Astute) is indicated in the figure by the area labelled 'other UK technical work'. It includes demands for the Future Aircraft Carrier (CVF), the Mari-

---

[11] The estimate of other UK technical work is from Mark V. Arena, Hans Pung, Cynthia R. Cook, Jefferson P. Marquis, Jessie Riposo, and Gordon T. Lee, *The United Kingdom's Naval Shipbuilding Industrial Base: The Next Fifteen Years*, Santa Monica, Calif., USA: RAND Corporation, MG-294-MOD, 2005. (Note the difference in scale from Figure 2.6.)

time Reach and Sustainability (MARS) programme, and the Type 45. Clearly, the demand for SSBN design resources would be ramping up at the same time as demand for other programmes taken together. Perhaps a third of the design force needed for the follow-on SSBN could move over from the Astute programme. The remainder needed would have to be hired from outside the current design workforce.

At the other extreme, the Vanguard class could have an operational life as long as 40 years, because improvements made during the recent mid-life refuelling and refit allow for improved reactor core life. There is precedent to such life extension in the decision by the United States to increase the planned operational life of the Ohio class of SSBNs to 40 years. Of course, further studies on the life of the hull structure would be required, and the combat and weapon systems would need upgrades to incorporate new technologies and to respond to emerging threats.

Figure 2.7
Design Effort for 25-Year Vanguard Class Life Coincides with Other Design Programmes

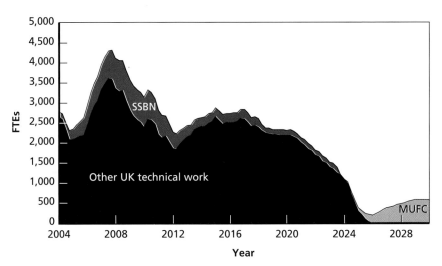

Extending the Vanguard class's life to 40 years would raise problems for sustaining the design base that are the opposite of those for a 25-year life. Figure 2.8 shows the relationship of follow-on SSBN design demands assuming a 40-year Vanguard-class life with design period for the MUFC programme. Here, the overlap in design efforts would be between the follow-on SSBN and the MUFC. However, the SSBN design demand drawdown is about evenly matched by the MUFC design demand run-up. There would be plenty of opportunity to put current-contract design resources to work on evolutionary development for the next contract—too much opportunity, in fact. In the 40-year-life case, a large gap appears *before* the SSBN follow-on design must get under way. Design demand drops below 200 FTEs for more than a decade. This gap, between current-contract Astute design and that for the follow-on SSBN, would be long—too long for further Astute work to fill.

Figure 2.8
Future Submarine Design Demands, Assuming Latest Possible Start for Design of New SSBN

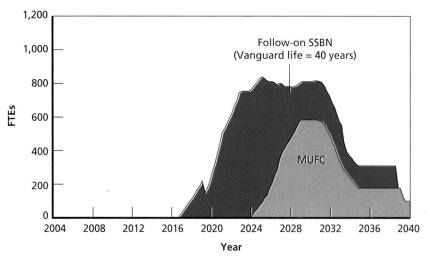

NOTE: Remaining Astute-related design work omitted to protect business-sensitive data.
RAND *MG326/1-2.8*

The advantage of the 40-year Vanguard class life is that the new submarine design effort would coincide well with future technical work. The run-up in SSBN design demand from about 2018 to 2023 would occur as the demand for other technical work is declining (see Figure 2.9). The portion of the SSBN workforce in excess of the submarine-specific core could thus come entirely from other programmes with no need to hire from outside the existing design workforce. This should allow a smooth transition of engineers and draughtsmen across programmes.

The potential advantages and disadvantages of the 25- and 40-year operational lives for the Vanguard class suggest a 30- or 35-year operational life would be the best option for sustaining the submarine design base. Indeed, a new submarine design effort starting in approximately 2008 appears attractive from an industrial base standpoint (see Figure 2.10). First, it provides time for additional development efforts for the Astute class between now and 2010. Engineers released from work on the first contract could staff such an effort.

Figure 2.9
Design Effort for 40-Year Vanguard Class Life Can Draw from Falloff in Demand from Other Programmes

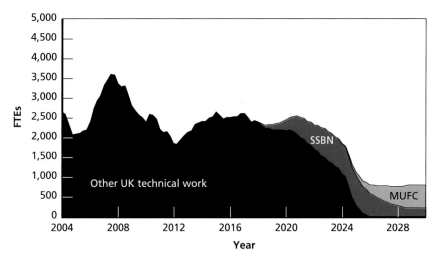

Figure 2.10
Future Submarine Design Demands, Assuming Start for Design of New SSBN to Replace Vanguard Class at Age 30

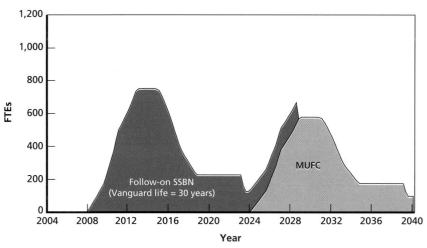

NOTE: Remaining Astute-related design work omitted to protect business-sensitive data.
RAND *MG326/1-2.10*

Figure 2.11
30-or-More-Year Vanguard Class Life Helps Fill 2010–2015 Trough in Demand

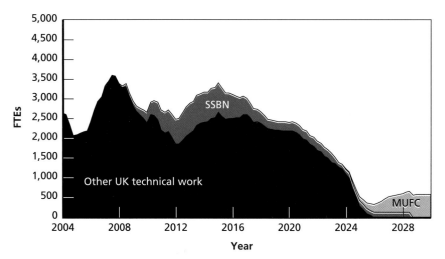

RAND *MG326/1-2.11*

Second, it provides a smooth transition between new design programmes. Finally, an SSBN design effort beginning in 2008 would help fill in the valley in the aggregate ship design demand profile between 2010 and 2015 (see Figure 2.11).

Sustaining a Design Core Between New Programmes

As shown above, even if the schedule of future design programmes is optimised for sustaining the design base, there will still be at least short periods of demand insufficient to sustain the core skills required over the long run. How will those in our proposed design core exercise and enhance their skills when there is no active design programme under way? Addressing ways to sustain the core is timely given that the Astute design team at Barrow is currently dropping towards the desired minimum 200-person level. Vendors will simultaneously be having problems sustaining their design resources.

Several options were identified as a way to sustain the core between new submarine design programmes. These options are not mutually exclusive; some could supplement or facilitate others. The options are as follows:

- spiral development of the Astute class
- continuous conceptual designs
- design of unmanned undersea vehicles (UUVs)
- design of diesel submarines for export
- collaboration with the United States or other countries
- collaboration within the United Kingdom.

As these activities do not fall within the scope of new-submarine design contracts, the MOD will need to allocate some funds to supporting the core. We estimate the requisite annual funds at roughly £15 million. Although not specifically oriented to the design of a specific submarine class, the core's activities should still benefit the latter.

Most of the options listed above could be planned to aid in incorporating new technologies and to reduce the production and ownership cost of the Astute class specifically and of nuclear submarines in general.

Regardless of the option or options chosen, BAE Systems Submarine Division at Barrow should manage the nuclear submarine design core because of their lead role in submarine design and construction. The majority of designers and engineers would come from Barrow. The designers and engineers from the vendors could relocate to Barrow or could interact with the rest of the core in a virtual environment with occasional group meetings to discuss issues and progress on design programmes. We will discuss later how this core could collaborate not only with design resources in other countries but also with those designers and engineers that support in-service submarines.

Spiral Development of the Astute Class

Spiral, or continuous, development is a commonly used practice to enhance platform performance by incorporating new technologies that become available or are needed to respond to new threats or missions. Spiral development efforts can also examine ways to reduce costs through incorporating manufacturing improvements or lessons learned. It is an effective method to sustain design resources between new programmes.

Long ship-operating lives and the extended construction periods for a given class of ships make spiral development almost a necessity. For example, Barrow may be building Astute-class submarines for 15 to 20 years. Threats and mission emphasis are likely to change over that period, and certainly electronic technologies will make significant advances. Without spiral development, by the time the final Astute-class submarines take their place in the force, their design will be 20 years old.

Two examples of the use of spiral development are the Los Angeles class of nuclear submarines and the DDG-51 class of destroyers in the United States. The Los Angeles class included three separate variants of submarines built over a 20-year period.[12] The spiral development efforts were able to sustain several hundred engineers and designers at Electric Boat and Newport News Shipbuilding. Likewise, the DDG-51 class included three variants of ships over a 20-year period that sustained a few hundred designers and engineers at Bath Iron Works and Ingalls Shipbuilding.

In the United Kingdom, most spiral development work is accomplished by the Submarine Support IPT within the MOD and by the contractors and shipyards that provide the in-service support to the nuclear submarine fleet. The engineers and designers at Barrow have typically played little or no role in design efforts after a submarine was delivered. The fact that BAE Systems now has design authority for the Astute class should result in a larger role for Barrow throughout the life of the class (unless the MOD assumes design authority on delivery). Barrow engineers and designers will most likely share Astute spiral development efforts with other organisations, such as the MOD, BES, and DML, but spiral development should provide work for part of the core submarine design team.

As mentioned, a worthwhile goal for spiral development efforts would be reducing the construction cost of the submarines. Barrow has recently returned to building submarines after a gap of several years and has been interacting with Electric Boat on the Astute-class design for the past two years. Certainly, lessons have been learned from these experiences that could suggest money-saving improvements in construction techniques.

The downside to spiral development as a core-sustaining device is that the majority of spiral development efforts are geared to updating combat system technologies, communications, or electrical

---

[12] Each variant marks a major upgrade in the class from previous ships in the class. For example, SSNs 719 through 750, the second flight of the Los Angeles class, have 12 vertical launch tubes for the Tomahawk missile that were not part of the first flight (SSNs 688 through 718). Also, SSNs 751 through 773, the third flight, modified the second flight by adding an improved sonar system and the ability to lay mines.

power distribution systems. Rarely does spiral development address basic hull forms or mechanical arrangements.

Continuous Conceptual Designs

Conceptual design efforts can complement spiral development. Whereas spiral development exercises skills related to combat systems, communications, and the electrical power distribution system, conceptual design studies encourage skills application and innovative thinking regarding the hull, mechanical, and electrical power generation systems. Submarine designers should constantly be seeking new approaches and technologies, so some level of conceptual design activity should occur on a continuous basis.[13] If the decision is made to have a follow-on to the Vanguard class, then conceptual design studies should begin as soon as possible (for Initial Gate in 2008; see previous discussion of timing).

Some of the outputs of the conceptual designs may advance to the preliminary design stage. Most of these designs will advance no further (i.e., nothing will be built to the design), but one or more of the conceptual designs will lead to the next new class of nuclear submarines. Germany, for example, funds new submarine design efforts every six years; many of the resulting designs do not advance to the construction stage.

The main disadvantage of continuous conceptual design work is that it will sustain only a small number of nuclear submarine designers and engineers. Conceptual study teams are typically small, on the order of 25 to 50 members. There are other disadvantages of designing submarines without building to the design:

- the lack of the design discipline and risk management that accompany awareness that the design is to be built

---

[13] The MOD should also be involved in conceptual design activities, playing a lead role in funding and directing studies and in disseminating promising results to key decisionmakers. See companion document, *The United Kingdom's Nuclear Submarine Industrial Base, Volume 2: MOD Roles and Required Technical Resources*, MG-326/2-MOD, forthcoming.

- the lack of feedback from lessons learned in construction to the design process
- the inability to judge the quality (and feasibility) of the design without actually building and testing it.

Design of Unmanned Undersea Vehicles

In the past few years, there has been an increased interest in unmanned undersea vehicles for both military and commercial uses.[14] Outside the military world, there are approximately 500 UUVs and autonomous undersea vehicles (AUVs) currently in use for functions such as marine science and oceanography, marine resource exploration and exploitation, hydrographic survey, and environmental protection. In the defence sector, UUVs are viewed as complements to manned systems in areas such as intelligence, surveillance, reconnaissance, communications, anti-submarine warfare, and mine countermeasures.

The United Kingdom has invested significant research funds into understanding the potential of UUVs. For example, for the Marlin programme, BAE Systems is developing an electrically propelled UUV. The programme's goal is to demonstrate the ability to launch and control UUVs from a manned submarine and to assess how submarine-launched UUVs could contribute to the Royal Navy's operations.[15]

UUVs will be vital components of future naval force structures, much as unmanned aircraft are becoming key components of air operations. Research and development work will continue, and that work can help sustain a small portion of the core of submarine designers and engineers between new nuclear submarine design programmes. The disadvantage of UUV designs for sustaining the core is that the majority of the design work is very different from the efforts

---

[14] A useful description of the worldwide research in UUVs is contained in 'Executive Overview', *Jane's Underwater Technology*, 12 January 2004. Also, the Southampton Oceanography Centre has hosted an annual conference on UUVs since 1999. See www.uuvs.net.

[15] See 'Mine Warfare: Underwater Vehicles, United Kingdom', *Jane's Underwater Systems*, 10 June 2004, for a description of the Marlin programme.

required for new nuclear submarines. UUV design efforts focus on miniaturisation and on electronics and communications. UUV propulsion systems will almost certainly be nonnuclear, and the design and engineering efforts geared towards crew-related issues do not arise in the design of UUVs. Even with these disadvantages, some UUV design work should be directed to sustaining the nuclear submarine design core.

Design of Diesel Submarines for Export

If design-only efforts could help sustain the nuclear submarine design and engineering core, it might seem that design to build for export might also have a role to play. This, however, is not an attractive option. Although the United Kingdom has recently sold its four Upholder-class submarines to Canada, the conventional submarine export market is characterised by several established sellers. Germany has had the most success in selling conventional submarines to other countries and would represent the strongest competition to a UK entry into the market. In addition to Germany, Russia has sold several submarines to countries such as China, Greece, and India. France, collaborating with Spain, has had considerable success with its Scorpene class. The Netherlands has developed the Moray submarine for export; although there have been no sales to date, several countries have expressed interest.

These sellers are competing for the business of relatively few buyers. China and India have purchased submarines in the past but are now turning to building boats of their own design. Otherwise, the demand side of the market has been limited to Pakistan (buying from France), Greece (from Germany and Russia), Chile (from the France-Spain partnership), and Indonesia (from Germany).[16]

---

[16] For a full discussion of the difficulties facing UK shipbuilders in entering the military export market, see John Birkler, Denis Rushworth, James Chiesa, Hans Pung, Mark V. Arena, and John F. Schank, *Differences Between Military and Commercial Shipbuilding: Implications for the United Kingdom's Ministry of Defence*, Santa Monica, Calif., USA: RAND Corporation, MG-236-MOD, 2005.

An additional problem facing UK submarine designers is developing an inexpensive submarine for export. The safety considerations associated with nuclear propulsion and the more stringent operational environments of UK submarines result in a high-end mind-set that may make any export design too complex and costly.

Collaboration with the United States or Other Countries

The United Kingdom is not alone in facing the problem of how to sustain nuclear submarine design and engineering skills between new programmes. The United States, for example, is confronting a similar problem as the design effort for the Virginia class is ending and the next new programme is a decade or more in the future.

Collaboration in nuclear submarine design would involve various designers and engineers from one country working with a design team in the other country, either on a new submarine design programme, on long-range conceptual studies, or on other design activities. For example, the United Kingdom is collaborating with the United States on the Astute-class programme as Electric Boat is providing various personnel and services to Barrow to help in finalising the design drawings.

There are a number of advantages of collaboration for both countries. It can help sustain the design core in both countries by providing meaningful work during gaps in each country's submarine design programmes. This would reduce the number of designers and engineers engaged in less-productive work or on overhead when no design programmes are under way. Collaborative efforts could also provide a cadre of designers and engineers from outside a country to help build the needed design resources, that is, to be part of the growth portion of a new design team (see Figure 2.4). Collaborative efforts could draw on the best nuclear submarine design skills from both countries and inject new ideas and methods into the design process. Finally, collaboration would help aid the interoperability of the two countries' submarine forces.

Collaboration also has a number of disadvantages. The new design programmes in each country would have to be coordinated so that they occur sequentially versus concurrently. Also, collaboration on systems as complex and sensitive as nuclear submarines would require a high degree of technical interchange and a sharing, to some degree, of proprietary, intellectual, and classified information. In fact, establishing the ground rules and boundaries for sharing sensitive information may be the biggest hurdle to overcome in a collaborative environment. Such sharing would require not only the formulation of working arrangements between the two governments but also cross-national cooperation between private submarine design and construction firms. It may also be necessary to invest in common design tools.

Collaborative Models. Collaboration could work in a number of ways. Figure 2.12 shows five possible models and the relative staffing levels for each model. The right end of the figure represents the current practice in the United Kingdom and the United States—doing all design work within the country. This approach requires a full staff of designers and engineers and provides a high degree of technical security.

At the left end of the figure, the other extreme, a country basically buys the design for a new submarine. Such a situation would exist if the United Kingdom bought design plans for the Virginia class and built to print in a UK shipyard. That would represent a near

Figure 2.12
Potential Collaborative Models

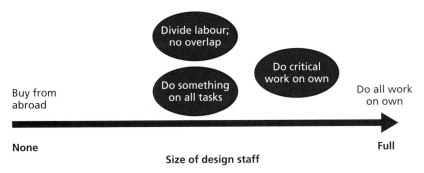

total divestiture of nuclear submarine design resources.[17] The potential problems with this option include

- issues of technology interchange
- the remaining requirement to maintain nuclear power plant design resources (assuming the purchased design did not include the nuclear propulsion system)
- the requirement for engineers to support the in-service submarines
- risks to the vendor base if the purchased design requires materiel and equipment from another country's vendors
- the possibility that the design available for purchase is for a submarine too elaborate and expensive to build.

The three models shown between these two extremes in Figure 2.12 require staffing levels that lie between no staffing and full staffing. One model ('divide labour; no overlap') envisions each country maintaining a set of design capabilities with no duplication. That is, neither country would have a complete set of skills but rather the total set of various design and engineering skills required for a new programme would be divided between the two. In this model, each country would have specific skills in depth, but neither would have the breadth of all skills required. When one country starts a new design programme, the other country would provide its skilled people to the programme. This model would allow specialisation by each country in specific areas. However, it would make it very difficult to revert to a self-sufficient model.

A second model ('do something on all tasks') would require each country to maintain all skills at a low level. Here, there would be breadth in each country but not depth. When a new design programme commences in one country, the other country provides

---

[17] It would not be a complete divestiture because the United Kingdom would still have to maintain some level of design and engineering resources to support the construction programme and the in-service submarines.

resources of all skills. This model would allow a country to revert to complete design capabilities by augmenting each of its skill bases.

The final model ('do critical work on own') combines the features of the previous two models. Each country would maintain all skills at a low level and certain critical skills at full capability. In many ways, this is the model currently in practice between Electric Boat and Barrow, in which Electric Boat is providing certain design resources to augment Barrow's capabilities. Here, in contrast to the other models, some overlap of programmes is possible. It is also the easiest from which to revert to fully self-sufficient design. However, this model requires greater staffing levels than the previous two.

Issues to Resolve in All Models. As mentioned previously, the most important issue to resolve for any collaborative model is establishing the intellectual boundaries and facilitating information exchange. Both countries' governments as well as the private firms involved with submarine design and construction in each country must address this issue. A collaborative model would not necessarily require full sharing in all highly sensitive areas. For example, each country could maintain its nuclear-propulsion system designs. But without a large degree of information sharing, collaborative models could fall short of desired goals.

Other issues include whether collaborative teams should be collocated and whether common design tools are required. For long-range conceptual studies, teams could work largely in a virtual environment. Some relocation of one country's designers and engineers to the other country may be needed to facilitate interactions and to allow indoctrination into the other country's methods and processes. A larger degree of collocation would be needed for the preliminary and detailed design portions of new submarine programmes. For example, a number of Electric Boat engineers and designers have relocated to Barrow to assist with the detailed design of the Astute-class programme.

Those involved in collaborative conceptual design should use a common computer design tool. Conceptual designs could then be ported to whatever design package each country uses for preliminary and detailed design. Collaboration during these later phases would also benefit from a common design tool. In the absence of one, collaborating designers and engineers would need to be familiar with both.

Finally, in addition to the submarine designers and engineers, design staff from the major vendors should also be part of any collaborative model. As with the shipyard designers, vendors from each country could share new ideas and approaches, and interactions could help to sustain the vendor base in each country.

Potential Model for Collaboration with the United States.[18] We recommend a collaborative model in which each country maintains all skills at a low level and certain critical skills at full capability. Collaboration could start small with up to 50 designers and engineers from each country assigned to work on one team. Neither the United Kingdom nor the United States is currently involved in new submarine design programmes. However, each country is in the initial stages of producing a new attack submarine class (Astute and Virginia, respectively), and each may start planning for its next ballistic missile submarine programme if future decisions result in the need for a new SSBN class. Approximately half the collaborative team could focus on conceptual designs of new SSBNs while the other half focuses on spiral development of the current production submarines. That is, a number of Barrow engineers could be assigned to Electric Boat to help with spiral development of the Virginia class and on conceptual studies for a new US SSBN. Likewise, Electric Boat engineers could be assigned to Barrow to participate in the spiral development of the Astute class and on conceptual studies for a new UK SSBN if the United Kingdom decides to produce a follow-on to the Vanguard class (or a new UK SSN if there is no follow-on to the Vanguard class). Such a plan would allow each country to take advantage of what the other has learned, and it would get the collaborative process off the ground. Because acquisition programmes are running

---

[18] We consider collaboration with the United States on the assumption that the nuclear agreements in place between the United States and United Kingdom would preclude collaboration with a third country on nuclear submarine designs. Also, there are very few other potential partners.

concurrently in the two countries, however, the plan would not materially aid in sustaining a submarine design core in either country.

From this start, the collaborative model could grow to allow the majority of the 200 UK core designers and engineers to work with the United States on new submarine programmes. The core of UK submarine design resources would be assigned to a US design programme, and US designers and engineers would be assigned to the next new UK nuclear design programme.

Collaboration Within the United Kingdom

Collaboration between organisations within the United Kingdom can also offer advantages and should be seriously considered. Such intra-UK collaboration could take either (or both) of two forms: collaboration between organisations that design and construct submarines and those that provide in-service support, and collaboration between the various shipyards and other organisations that design and build surface ships or submarines.

Currently, DML and BES provide support to in-service submarines, including design and engineering services. These two organisations team with the Submarine Support IPT in the MOD to develop maintenance schedules and plans and the supporting work packages for the Swiftsure-, Trafalgar-, and Vanguard-class submarines. They also manage the facilities and operations at the Devonport Dockyard and Faslane Naval Base, respectively.

The design and engineering resources at Barrow play little if any role in the support of in-service submarines, since design authority for the current classes of submarines rests within the MOD. However, design authority for the Astute class is currently held by BAE Systems. If BAE Systems maintains this authority once the Astute-class submarines enter the fleet, the design team at Barrow must interact with DML, BES, and the Submarine Support IPT. This would require the various organisations to collaborate on the design and engineering activities needed to support in-service submarines.

Even if design authority for the Astute class reverts to the MOD once the submarines enter service, collaboration among Barrow, DML, BES, and the MOD would still be desirable. Spiral development efforts for the Astute class should take advantage of lessons learned during the support of all classes of in-service submarines. These lessons learned could help incorporate changes in the Astute-class design that reduce maintenance and other through-life costs. Also, DML and BES, which maintain in-service submarines, would require the design and construction knowledge of the Barrow engineers to adequately plan maintenance work packages. Finally, and most importantly to the topic at hand, participating in the support of in-service submarines would help sustain a part of the nuclear submarine design core.

Concerning collaboration among the organisations involved in the design of ships for the MOD,[19] there are currently a number of new shipbuilding programmes under way or about to start in the United Kingdom. The short-term demand for design resources may exceed the available supply. Even so, most new shipbuilding programmes in the United Kingdom already involve teaming of two or more shipbuilders. BAE Systems Naval Ships Division and VT Shipbuilding are teaming on the design and construction of the Type 45 destroyers. BAE Systems, VT Shipbuilding, Swan Hunter, and BES Rosyth will possibly collaborate on the design and construction of the CVF ships. Two or more firms will design and build the MARS ships. It is questionable whether one shipbuilder currently has adequate resources to conduct a ship design programme by itself.

Once the new destroyers, submarines, carriers, and auxiliary ships are designed, there will be few if any new programmes to place demands on the shipbuilding design base. All shipbuilding organisations will face challenges in sustaining their design resources. The various organisations might need to collaborate even more broadly than they are now doing to share scarce design resources for the few new programmes that will begin after the next decade.

There are questions of how to organise and manage a UK-wide collaborative design team. Company boundaries and interests would have to give way to more nationally oriented concerns; MOD roles

---

[19] In addition to the shipbuilders, the collaboration should include private design firms like British Maritime Technology and knowledgeable organizations like Qinetiq.

would have to be carefully defined. Competition for new designs might not be feasible. Rather, new design programmes may be assigned to the national team. Competition, if it is possible at all, may be limited to new construction. These types of issues will require careful consideration, but some form of a nationwide design consortium may be required in the future.

Summary

The unique nature of many submarine design and engineering skills requires careful management of the nuclear submarine design resources. If the United Kingdom wishes to keep the option open for future SSN or SSBN programmes, then it must retain a core of approximately 200 nuclear submarine designers, engineers, and draughtsmen. This core would form the foundation of the team required for any new submarine design programme. The core could be sustained through multiple actions including the spiral design of submarines currently in production, continuous conceptual studies, and designs of new UUVs. Collaboration with another country such as the United States could also help sustain the design core and would foster the exchange of ideas and greater interoperability of the two country's submarine forces. Issues surrounding collaborative efforts would require close coordination among the governments of the two countries as well as among the private firms involved in nuclear submarine design and construction.

There are many issues that play in sustaining nuclear submarine design resources, not the least of which is whether the United Kingdom will retain a submarine-based deterrent. If the decision is made to replace the Vanguard class and if the Vanguard class holds to its original 25-year operational life, the design programme would need to begin immediately. That would provide sufficient demands to sustain the nuclear submarine design base but would leave little or no time for continuing development of the Astute class and would create a long gap in design efforts until the MUFC class starts. Also, demands for designers and engineers to support the Type 45, CVF, and MARS programmes will compete for design resources with a follow-on to the Vanguard class.

At the other extreme, the operational life of the Vanguard class could possibly be extended to around 40 years. This would allow time for spiral development of the Astute class, but it is unlikely that the Astute class could fill the large gap in demand that would open prior to commencement of new SSBN design. The new SSBN design programme could well overlap the new design programme for the MUFC. Currently, there are insufficient submarine design resources in the United Kingdom to support two concurrent programmes.

From the perspective of sustaining nuclear submarine design resources, an operational life of slightly more than 30 years for the Vanguard class is preferred. This allows spiral development of the Astute class and minimises the gap prior to the design effort for the MUFC.

CHAPTER THREE

Maintaining Nuclear Submarine Production Resources

BAE Systems Submarine Division's shipyard in Barrow-in-Furness is the sole producer of nuclear submarines in the United Kingdom. Barrow has a long and distinguished shipbuilding history: It has not only built submarines for more than 100 years but has designed and constructed many first-of-class surface ships, including the Type 42–class destroyers and the Invincible-class aircraft carriers. Barrow has recently delivered two Wave-class auxiliary oilers and two Albion-class landing platform, docks (LPDs). These four surface ships helped fill the production gap at the shipyard between the end of the construction of the Vanguard-class SSBNs and the start of the current Astute-class SSNs.

The workforce levels at Barrow have dropped dramatically over the past decade. Currently, there are approximately 3,000 white-collar and blue-collar employees at Barrow, fewer than one-quarter of its more than 13,000 employees in 1990. The Barrow shipyard has also had a number of changes in ownership over the past several years. After almost a century of ownership by Vickers Shipbuilding and Engineering Limited (VSEL), GEC Marconi acquired Barrow in 1995. The shipyard again changed hands in 1999 when BAE Systems bought it.

Importance of Continuous Production

In the early part of the past decade, the US Navy faced a major decision concerning its nuclear submarine industrial base. At the request of the Office of the Secretary of Defense, RAND evaluated the costs and risks of shutting down Electric Boat for a number of years and then restarting when new submarine construction was required.[1] The costs of such a strategy were estimated at approximately £1 billion, with the majority of the costs apportioned to rebuilding and retraining the workforce. Over and above the costs involved, reopening a nuclear submarine shipyard after a period of dormancy would involve a number of nonmonetary risks, including

- the possibility that the shipbuilder, and the vendors that support nuclear submarine construction, would have no desire to reenter the business
- the potential for accidents, especially a nuclear accident, due to the inexperienced workforce
- the possibility that the public would not allow nuclear activities to resume.

Given the costs and risks, RAND recommended that an additional Seawolf-class submarine be constructed to bridge the gap in nuclear submarine production at Electric Boat. Based on the recommendation, a third Seawolf-class submarine was constructed, the USS *Jimmy Carter*.

Although the costs and risks of shutting down Barrow and restarting in the future are likely different from those associated with the analysis of Electric Boat, we believe the cost, schedule, and risk associated with a shutdown/restart option are too high. Therefore, if the United Kingdom desires to maintain nuclear submarines in its force structure for the foreseeable future, the MOD must carefully manage new submarine procurement contracts to ensure that Barrow

---

[1] Birkler et al. (1994).

and the other organisations that comprise the nuclear submarine industrial base can efficiently build the boats.

It is possible, as was done between the end of the Vanguard class and the start of the Astute class, for submarine production to stop for some period of time as long as some relevant shipbuilding work is done at the shipyard. However, the Vanguard-Astute gap is one reason cited for the cost and schedule problems faced by the Astute programme. Although a gap in submarine production is possible, it is not preferred because it is very difficult for production workers to maintain nuclear qualifications without building nuclear-powered vessels.

The construction of nuclear submarines requires various skilled workers with suitably qualified and experienced personnel (SQEP) certification. Up through the Vanguard class, welders, electricians, pipe fitters, weapon systems workers, and team leaders had relatively high requirements to maintain their nuclear qualifications. Now, regulation has broadened the SQEP requirement to include any work that can impact the nuclear steam-raising plant.

Historically, SQEP status was gained and maintained through rigorous training. More recently, qualification has also required experience in nuclear-related work. That is, SQEP certification requires a documented ability to actually carry out a specific task rather than a general nuclear qualification. This requires some level of ongoing nuclear-related work to sustain nuclear-qualified production skills.

Currently, the only construction work in the shipyard is for the first three boats of the Astute class. An estimate of the remaining production workforce demand associated with these three boats is shown in Figure 3.1. The main question we address in this chapter is how to schedule future submarine construction starts to most efficiently use the resources at Barrow and at the vendors that support submarine construction. A long-term view of future submarine programmes is needed to effectively answer this question.

In the following sections, we show how such a long-term view frames interrelated decisions regarding production timing and fleet size for the future submarine classes and the remainder of the Astute

Figure 3.1
Current Production Workforce Demands at Barrow

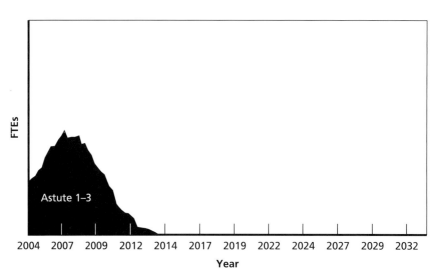

class. We assess the impacts of these decisions on the temporal profile of the demand for production labour and on labour costs. In separate exercises, we examine how long construction of Astute 4 might be delayed without letting the SSN fleet size drop below various levels as well as the options and costs associated with a gap in submarine production. We also discuss the filling of any residual production gaps at Barrow with non-submarine work. We conclude with an appraisal of the vitality of the vendor base supporting submarine production.

A Look at Future Programmes

Figure 3.2 shows one possible schematic representation of future nuclear submarine production. There is the current Astute contract for the first three boats plus three potential future programmes. A recent announcement by the Secretary of State for Defence has set

Figure 3.2
Current and Potential Future Nuclear Submarine Programmes

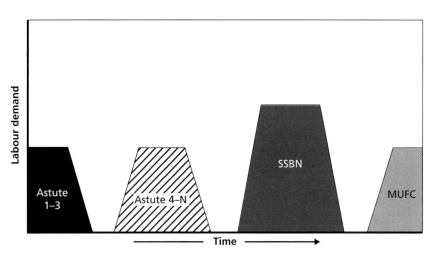

RAND *MG326/1-3.2*

SSN force size at eight submarines.[2] This suggests a need for five additional Astute-class submarines, which will replace the retiring Swiftsure- and Trafalgar-class boats. Although no decisions have been made as yet, if an underwater strategic nuclear deterrent is to be sustained, the Vanguard-class SSBNs will have to be replaced at some point in the future. Finally, a replacement SSN class will be required when the Astute-class boats reach the end of their operational life. The MUFC programme represents this potential replacement.[3]

We begin the analysis by initially assuming there will be a follow-on to the Vanguard class. For the most efficient use of the production base, the four programmes in Figure 3.2 should overlap in a way that allows smooth transition of production resources from one programme to the next (see Figure 3.3). To accomplish this, decisions

---

[2] See UK Ministry of Defence, 'Delivering Security in a Changing World: Future Capabilities', white paper presented to Parliament by the Secretary of State for Defence, July 2004.

[3] The follow-on SSBN and possibly the MUFC may not be entirely new classes of submarines but may be a variant of the Astute design.

Figure 3.3
Overlap of Programmes Should Provide Efficient Use of Production
Resources

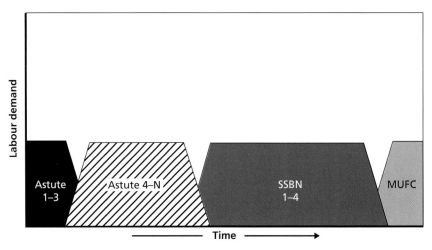

are needed on the start and end dates for each programme, the number of boats built, and the 'drumbeat'.[4] These various decisions are related; for example, the drumbeat is a function of the start and end dates and the number of boats in the class.

It is difficult to address these multiple decisions simultaneously. Therefore, we start by fixing one of the decision points and measure the resulting impact on the other decisions. In doing this, we highlight the relationships between the various decisions and provide an analysis framework that is flexible enough to help MOD decision-makers understand the implications of various policies. There are uncertainties in many of the factors used in the analysis, such as the operational lives of the current and future submarines, the desired force structures, and the build periods for new submarines. The

---

[4] We use the term *drumbeat* throughout this report to represent a consistent production rate. An 18-month drumbeat suggests the construction of a new submarine begins every 18 months.

resulting analysis should, therefore, be viewed as representative of the impact of various decisions, not as definitive predictions of future events.

As in the design analysis, we begin by assuming the MUFC will be a class of submarines and by fixing the start of its construction. Again, the first MUFC boat should be delivered in 2034. Assuming a seven-year build period for the first of class, the first MUFC should start construction in the first quarter of 2027. With this starting point, we next turn to how best to schedule production of the follow-on SSBN class to provide efficient use of production resources.

Timing of Production for the Next SSBN Class

If the Vanguard class is held to its originally planned operational life of 25 years, the first of class will leave the force structure in 2018. Assuming an eight-year build period for the first boat of the next class, construction would have to begin in 2010. The impact on the demand for FTE production resources at Barrow is shown in Figure 3.4. The workload demand for the first three Astute-class boats is based on the Barrow estimate for completion of the current contract (see Figure 3.1). We use the projected hours for the third boat as an estimate of workload for future SSNs (including the MUFC). Finally, we assume that an SSBN would require 50 percent more hours to build than an SSN and that four SSBNs will be built.

Two problems from the production base perspective exist with a 25-year Vanguard class life. First, there is little time to produce additional boats of the Astute class before the SSBN programme begins seriously competing for production resources. Scheduled retirements of the current in-service submarines will lead to very low SSN force structure levels if additional Astute-class submarines are not built in the next decade. Second, there is a substantial gap between the end of the follow-on SSBN production and the start of the MUFC production.

Figure 3.4
Projected Workforce Demands at Barrow for a 25-Year Vanguard Class Life

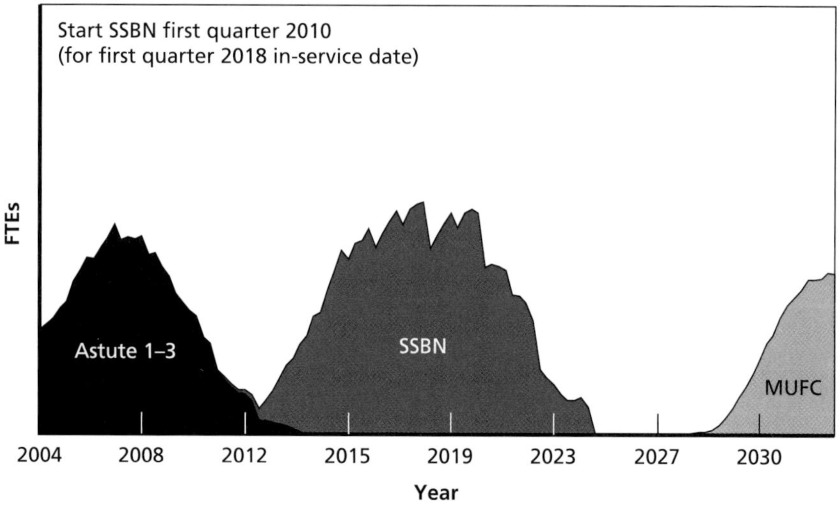

Figure 3.5 shows the impact on the demand for production resources of a 40-year Vanguard class operational life. The problems here are perhaps even worse. Although there is plenty of opportunity to finish the Astute programme unimpeded, there would be a long gap between the production of the last Astute-class boat and the start of the SSBN follow-on programme. Furthermore, the coincidence of production between the MUFC and the follow-on SSBN boats would require a daunting increase in production resources.

To maximise production base efficiency, we should begin by setting an optimal overlap between the SSBN and MUFC programmes to ensure a smooth transition. The SSBN drumbeat and construction duration (assumed to be eight years) would then determine when the programme should start and when the Vanguard class should begin to retire. Based on the experiences of previous UK and US build programmes, we infer that an approximately five-year overlap would provide the smoothest workforce transition between the

Figure 3.5
Projected Workforce Demands at Barrow for a 40-Year Vanguard Class Life

end of one programme and the beginning of the next. With the MUFC starting construction in 2027, the efficient use of production resources suggests that the follow-on SSBN programme should end construction in 2032. Table 3.1 shows, in each cell, when the first SSBN boat must start construction and the range of ages of the Vanguard-class boats when they are replaced,[5] given the specified combination of SSBN drumbeat and fleet size. (Although we have so far been assuming a fleet size of four boats, we allow here for the possibility of a three-boat fleet.)

For a future force of four SSBNs, an efficient transition to MUFC construction and reasonable drumbeats of 24 to 36 months suggest a Vanguard-class operational life of between 30 and 34 years.

---

[5] We assume a Vanguard-class boat is replaced when a follow-on SSBN boat is delivered. Because the Vanguard-class boats were not delivered to a set drumbeat, the age of each Vanguard-class boat varies at retirement.

Table 3.1
Start Dates for a Follow-On SSBN Based on Drumbeats and Force Levels

SSBN Force Size	SSBN Drumbeat		
	24 Months	30 Months	36 Months
4	First quarter 2018 (32 to 34)	Third quarter 2016 (31 to 33)	First quarter 2015 (30 to 33)
3	First quarter 2020 (34 to 36)	First quarter 2019 (34 to 36)	First quarter 2018 (33 to 36)

NOTE: Each cell shows production start of first of class of follow-on SSBN programme and in parentheses the range of operational lives of Vanguard-class boats.

This result is consistent with the analysis of sustaining design resources. If a force structure of only three future SSBNs is desired, then the operational life of the Vanguard class increases by a few years.

Given the follow-on SSBN start dates and drumbeats, we can place a labour-demand profile on the timeline. Figures 3.6 and 3.7 show the profiles for a four-ship SSBN programme with drumbeats of 24 months and 36 months, respectively. The quicker SSBN drumbeat results in a slightly higher peak demand for production workers than is required for the following MUFC programme. The slower drumbeat[6] provides an almost uniform demand for production workers at Barrow as the workforce transitions from the follow-on SSBN to MUFC production.

Fleet Size and Production Timing for the Remaining Astute-Class Submarines

We have shown how SSBN fleet size and drumbeat determine a start date for SSBN construction. SSBN construction will be preceded by

---

[6] A drumbeat that is too slow may result in inefficiencies at the skill level. Such potential skill-related problems are discussed later in this report.

Figure 3.6
Demand for Production Resources at Barrow from a 24-Month Follow-On SSBN Drumbeat

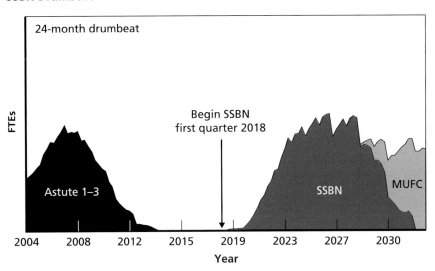

Figure 3.7
Demand for Production Resources at Barrow from a 36-Month Follow-On SSBN Drumbeat

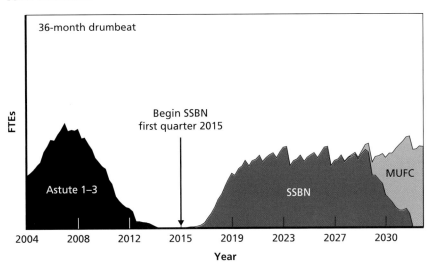

production of some number of Astute-class boats above the three currently contracted. (That number has been set at eight, but that could change.) Let us again assume a five-year overlap between programmes (Astute and follow-on SSBN in this case) and allow nothing but the typical drumbeat between Astute 3 and Astute 4. The production valley shown in Figures 3.6 and 3.7 can then be filled with Astute-class submarines based on a desired fleet size and production drumbeat. Because the Astute programme end date is determined by SSBN drumbeat, given an SSBN fleet size, the Astute fleet size achievable is a function of SSBN and SSN drumbeat. Table 3.2 shows the number of Astute-class boats (including the first three currently on contract) that could be built for different SSN and SSBN drumbeats, assuming four new SSBNs are desired.[7] Each cell in the table shows the Astute fleet size and the start date for the fourth Astute-class boat.[8]

Table 3.2 also shows there are different sets of SSN and SSBN drumbeats that produce the same Astute fleet size. The slower the SSBNs are built, the less time is available to build SSNs and the faster the SSNs must be built to yield the same number of boats. For example, a total of eight SSNs (the first three Astute-class submarines plus five additional boats) is possible with drumbeats of 30 months for the SSNs and 24 months for the SSBNs, 24 months for the SSNs and 30 months for the SSBNs, and 18 months for the SSNs and 36 months for the SSBNs.

The eventual fleet size is not, however, sustained over the transition between the Swiftsure and Trafalgar classes and the Astute class.

---

[7] The SSBN production intervals considered tend to be longer than those for the SSN, because the labour demand at any given time during SSBN production is typically higher than the labour demand during SSN production.

[8] Astute fleet size and the Astute 4 start date are determined by beginning at the desired overlap point of five years following the start of SSBN production. We then subtract the seven-year Astute construction period and tally other Astute starts periodically before that, where the period is the drumbeat. The process is finished when it is not possible to insert another Astute start without leaving too small an interval after Astute 3's start. The variations in Astute 4 start date are thus a result of the vagaries of that dovetailing with Astute 3 production.

Table 3.2
Number of Astute-Class Submarines for Various SSN and SSBN Drumbeats,
Assuming Four New SSBNs

SSN Drumbeat	SSBN Drumbeat		
	24 months (32 to 34)[a]	30 months (31 to 33)[a]	36 months (30 to 33)[a]
18 months	10 (first quarter 2007)	9 (first quarter 2007)	8 (first quarter 2007)
24 months	9 (first quarter 2006)	8 (third quarter 2006)	7 (first quarter 2007)
30 months	8 (first quarter 2006)	7 (first quarter 2007)	6 (first quarter 2008)

NOTE: Each cell shows the number of Astute-class boats plus the start of the fourth Astute-class submarine (in parentheses).
[a] Range of life for Vanguard class submarines.

Currently serving boats are coming out of the fleet at irregular intervals, so the fleet will sometimes be above the number shown in Table 3.2 and sometimes below it. Figure 3.8 shows how the combined SSN fleet size varies over time for the 30-month SSN and 24-month SSBN drumbeats. The SSN force structure drops over the next several years due to the retirement of the Swiftsure-class submarines and the early Trafalgar-class boats. Once the Astutes begin entering the fleet in 2009, the SSN force structure stays fairly steady at eight submarines, with occasional increases to nine or decreases to seven. These variations are due to the mismatch between the in-service dates of the various Trafalgar-class boats (and therefore their retirement) and the introduction of Astute-class submarines.

For the 24-month SSN and 30-month SSBN drumbeat, the pattern is similar (Figure 3.9). However, there is more time when there are nine SSNs in the force and less time when there are seven compared with the previous case because of the quicker SSN drumbeat. This is all the more so when the SSN drumbeat is dropped to 18 months (Figure 3.10). (These patterns are almost entirely a function of the SSN drumbeat. The only effect of the SSBN drumbeat is

Figure 3.8
Number of SSNs in the Fleet (30-month SSN/24-month SSBN Drumbeats)

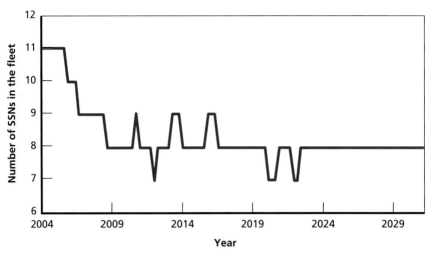

Figure 3.9
Number of SSNs in the Fleet (24-Month SSN/30-Month SSBN Drumbeats)

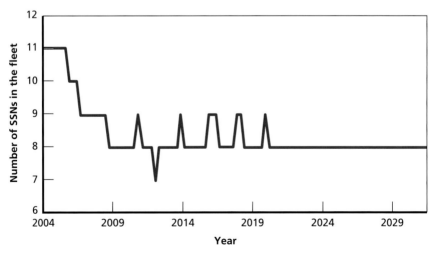

Figure 3.10
Number of SSNs in the Fleet (18-Month SSN/36-Month SSBN Drumbeats)

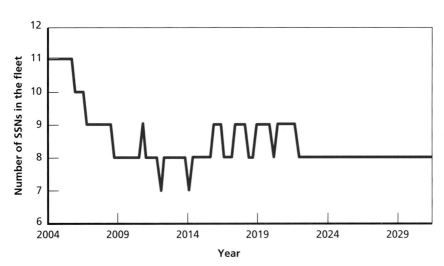

through its interaction with the SSN drumbeat to move the Astute 4 start date up or back as much as a year.)

The relative rates at which SSNs and SSBNs are built have implications for the workforce-demand profile. Figure 3.11 shows the workforce demand at Barrow for the 30-month SSN/24-month SSBN case. The combination of a slow SSN drumbeat and a faster SSBN drumbeat results in a drop in demand for SSN labour following Astute 3 and then an increase in workforce demand for the SSBNs (which have greater workloads per boat compared with the SSNs). If eight SSNs are desired, this particular combination of drumbeats may not be the best to avoid problems in managing the workforce at Barrow.

If the SSNs are built more quickly and the SSBNs more slowly, the demand for production workers at Barrow smoothes out (see the

Figure 3.11
Workforce Demands at Barrow (30-Month SSN/24-Month SSBN Drumbeats)

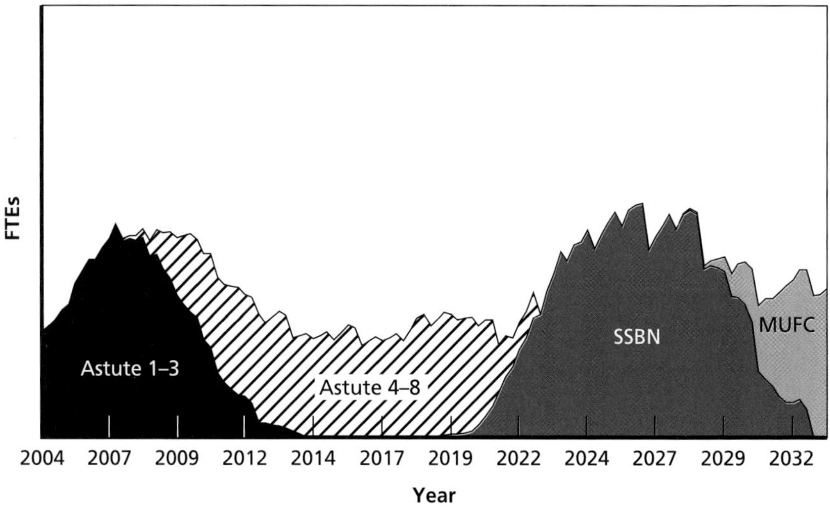

24-month SSN/30-month SSBN case in Figure 3.12). Workforce levels drop slightly during the build of the additional five Astute-class submarines and then rise for the build of the SSBN boats, but the variation is much less than in the previous case.

Figure 3.13 shows the workforce demands at Barrow for the 18-month SSN/36-month SSBN drumbeats. This option produces the most uniform demand for production workers at Barrow.

Table 3.3 (p. 56) compares indicative quantities from the six graphics. The table shows, for each drumbeat combination, the number of quarters over the 13-year transition period that the fleet size is projected to be over and under eight boats. It also shows, for each combination, the FTE peak as a percentage over the FTE minimum over the course of SSN and SSBN production.

There are obviously trade-offs when considering the combination of SSN and SSBN drumbeats that can produce a desired SSN force structure. Faster SSN drumbeats will introduce new SSNs to

Figure 3.12
Workforce Demands at Barrow (24-Month SSN/30-Month SSBN Drumbeats)

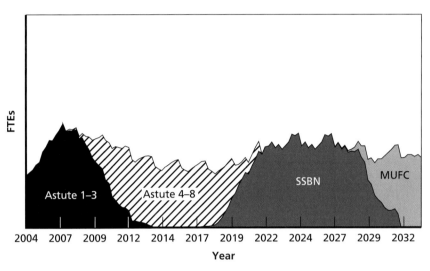

Figure 3.13
Workforce Demands at Barrow (18-Month SSN/36-Month SSBN Drumbeats)

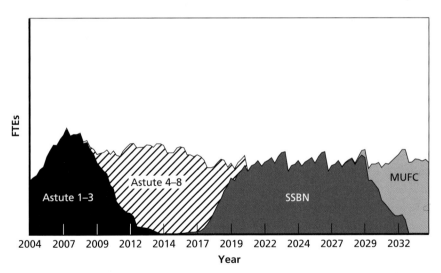

Table 3.3
Comparison of Fleet Size Variation and Workload Variation Across Different
SSN/SSBN Drumbeat Combinations

	—Faster SSN Construction—» «—Faster SSBN Construction—		
	30-Month SSN/ 24-Month SSBN	24-Month SSN/ 30-Month SSBN	18-Month SSN 36-Month SSBN
Quarters over/ under 8 boats in fleet[a]	7/6	8/1	19/2
FTE peak as percentage above FTE minimum[b]	160%	70%	45%

[a] After 2009; transition period is approximately 50 quarters long.
[b] SSBN peak as percentage above SSN minimum for first two columns; SSN peak as percentage above SSBN minimum for third column.

the force at a quicker rate resulting in higher force structures than desired for periods of time. Force structures can more closely match the desired level with slower drumbeats. However, slower SSN drumbeats reduce the workforce demands at Barrow and cause difficulties managing the workforce. Slower SSN drumbeats may also create problems with the various vendors that support submarine construction.

We conclude that, because workforce demands from an SSBN are greater than those for an SSN, the SSBN drumbeat should be slower than the SSN drumbeat to provide more uniform demands on the production workforce at Barrow. From that point of view, the preferable options for producing five additional Astute-class submarines are to set a 24-month SSN drumbeat followed by a 30-month SSBN drumbeat or to set an 18-month SSN drumbeat followed by a 36-month SSBN drumbeat.

Workforce Demands at the Skill Level

Examining the workforce implications of different build plans and different drumbeats at the total workforce level can mask problems that may exist at the level of specific skills. Submarine production involves construction skills such as steelworking, welding, and ship fitting early in the construction process when the hull cylinders are fabricated. Outfitting skills are used later in the process when the various electrical, piping, heating, ventilation, air conditioning, crew accommodations, and other systems are placed either in the hull cylinders or into the complete submarine structure.

It is important to verify that in smoothing out overall demand, we do not concentrate construction skills or outfitting skills within short periods. We would not expect that situation, because we have set the later phases of constructing one class to overlap with the early phases of the next. Figures 3.14 through 3.16 show the demands for construction and outfitting skills[9] for the case of five additional Astute-class submarines (in addition to the first three Astute-class boats, four SSBNs, and the first three MUFCs). Each figure shows the demands for a different SSN/SSBN drumbeat combination. As expected, the results for the two skill sets mirror those for the workload as a whole. The 18-month SSN drumbeat followed by a 36-month SSBN drumbeat provides the most uniform demand over time for construction skills and for outfitting skills.

Labour Costs of Various Options

Buildups and drawdowns of the production labour force can be costly. Can these costs be quantified? To find out, we ran the labour force model contained within the RAND shipbuilding and force structure analysis tool developed to support previous UK and US analyses of the shipbuilding industrial base.[10] The labour force model matches the supply of production labour to demands for it under certain conditions that constrain how quickly a workforce can expand or contract. In addition to direct wage rates, the model includes the hiring, training, and proficiency costs of expanding the workforce and

---

[9] In addition to construction and outfitting skills, submarine construction also makes smaller demands on technical, management, and support skills.

[10] Mark V. Arena, John F. Schank, and Megan Abbott, *The Shipbuilding & Force Structure Analysis Tool: A User's Guide*, Santa Monica, Calif., USA: RAND Corporation, MR-1743-NAVY, 2004.

Figure 3.14
Demands for Construction and Outfitting Skills: Five Additional Astute-Class
Submarines (30-Month SSN/24-Month SSBN Drumbeats)

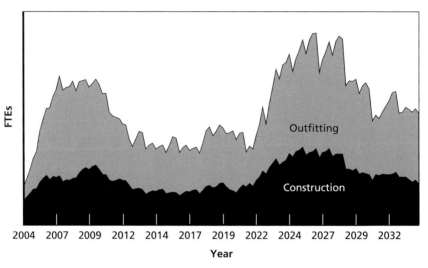

RAND *MG326/1-3.14*

Figure 3.15
Demands for Construction and Outfitting Skills: Five Additional Astute-Class
Submarines (24-Month SSN/30-Month SSBN Drumbeats)

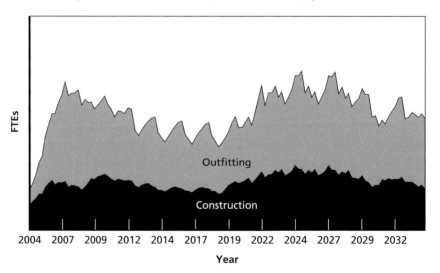

RAND *MG326/1-3.15*

Figure 3.16
Demand for Construction and Outfitting Skills: Five Additional Astute-Class
Submarines (18-Month SSN/36-Month SSBN Drumbeats)

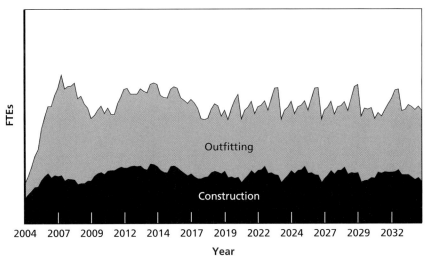

the termination costs of drawing down the workforce. Factors in the model are based on data and information received from Barrow plus data synthesised from a number of previous shipbuilding industrial base studies.

Table 3.4 shows the resulting estimates of the combined labour and overhead costs at Barrow for the SSNs built at different SSN and SSBN drumbeats. Each cell shows the number of additional Astute-class submarines built (beyond Astute 3) and our estimate of the direct labour and overhead costs per additional Astute class submarine at Barrow.

Although the values in Table 3.4 should not be viewed as definitive estimates, they suggest that of the three plans to produce five additional Astute-class submarines, the 18-month SSN/36-month SSBN drumbeat combination results, as might be expected, in the lowest cost. However, it is less than 10 percent below the cost of the

Table 3.4
Average Labour and Overhead Cost per Additional Astute-Class Submarine for Different SSN and SSBN Drumbeats (Millions of 2004 Pounds)

SSN Drumbeat	SSBN Drumbeat		
	24 months	30 months	36 months
18 months	7 £112	6 £113	5 £115
24 months	6 £121	5 £123	4 £125
30 months	5 £129	4 £132	3 £134

24-month SSN/30-month SSBN drumbeat plan. The cost advantage over the faster SSBN production rate is somewhat larger.

Delaying the Start of Astute 4

So far, we have been assuming no gap in submarine production at Barrow after the current contract ends with Astute 3. However, it may, for budgetary or other reasons, be desirable or necessary to suspend submarine production for a time. Here, we ask how long the next contract can be delayed and still keep the fleet size at a specified level[11] and what would be the effect on the labour-demand profile at Barrow. In concerning ourselves with these near-term dynamics, we will not thoroughly analyse the implications for SSBN and MUFC production.

The latest start dates for fleets of varying sizes can be represented in a fleet drawdown graph such as that in Figure 3.17. The line in the figure shows the projected UK SSN fleet size, given current plans for Swiftsure- and Trafalgar-class retirement and Astute 1, 2, and 3 in-

[11] Note that this differs from the previous analysis, in which the target fleet size could be missed from time to time until previous SSN classes had been fully retired. Here, we use an absolute minimum to determine the longest gap possible.

Figure 3.17
Start Dates to Maintain Various SSN Fleet Sizes

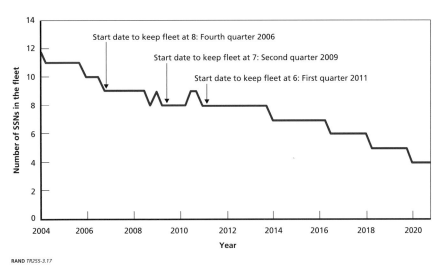

RAND *TR255-3.17*

service dates. The arrows indicate when construction of Astute 4 must start to maintain SSN fleet sizes of six, seven, or eight boats. The start dates are calculated straightforwardly. For example, for a desired SSN force size of seven boats, we note when the fleet would fall to six submarines and subtract seven years to allow for the replacement boat's construction.[12]

For an SSN fleet size of seven submarines, the start of construction for the fourth Astute-class boat can be delayed until mid-2009. For an SSN fleet size of six submarines, construction of Astute 4 does not have to begin until early 2011. Thus, it is possible to have a gap of several years at Barrow between the construction starts of the third and fourth submarines of the Astute class.

The potential workforce implications of the gap for an SSN force size of six submarines are shown in Figure 3.18. Total produc-

---

[12] Note that the date for eight submarines is a quarter earlier than the latest date given in Table 3.2; however, the latter plan did not keep the fleet size at eight or more (see Figure 3.10).

tion labour will fall as Astute 3 work winds down and will stay for three years at a level less than one-third of that eventually needed. We have already addressed the problems of such workforce fluctuations: loss of learning, subsequent production inefficiencies, and difficulties in maintaining certifications of key production personnel as suitably qualified and experienced. However, reductions in SSN force structure and budget priorities for other shipbuilding programmes could also lead to such a gap.[13]

Figure 3.18
Workforce Implications of Production Gap at Barrow for SSN Fleet Size of Six Submarines

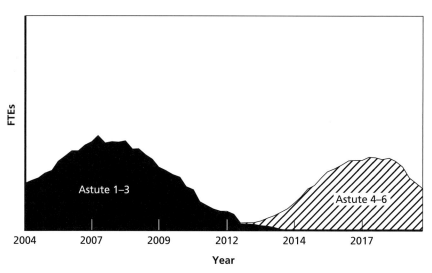

RAND MG326/1-3.18

---

[13] Note that the delayed SSN schedule would push SSBN production back, although there would probably still be enough time to produce a four-boat SSBN fleet at a 24-month drumbeat and still avoid too serious an overlap with MUFC production.

Implications of No Follow-On to the Vanguard Class

In the previous analyses, we assumed a follow-on SSBN to the Vanguard class. Here, we examine the options that the MOD may take if there is no follow-on to the Vanguard class. As mentioned earlier in this chapter, shutting down the Barrow shipyard for a period of time and then reopening it to resume submarine production would result in substantial costs and risks. Therefore, there should be some level of continuous production at the shipyard, preferably the construction of submarines. Given the current assumptions of eight SSNs in the fleet and a 25-year operational life, two strategies could help in achieving a continuous level of submarine production: lengthen the drumbeat of the remaining Astute-class boats and/or accelerate the start of the MUFC class.

Table 3.4 showed the cost penalty associated with longer SSN drumbeats of up to 30 months. A 36-month SSN drumbeat results in an additional cost penalty of approximately £10 million per boat or an average labour and overhead cost of approximately £140 million for the five remaining Astute boats (assuming a class size of eight).[14] The additional cost is due to the cost of labour turbulence (see Figure 3.19) and the increased overhead for the additional years of production.

In addition to the cost penalty, the 18-, 24-, and 30-month SSN drumbeats result in gaps of approximately 7, 5.5, and 4 years, respectively, between the end of the Astute class and the beginning of the MUFC. A gap of approximately two years occurs when the Astute-class drumbeat is extended to one new submarine start every 36 months. Assuming a five-year overlap is desired to reduce workforce turbulence, a 36-month drumbeat for five remaining Astute-class submarines suggests the in-service date for the MUFC should be

---

[14] This cost penalty does not include any vendor-related costs. A 36-month drumbeat may lead to substantial problems sustaining the vendor base, especially the nuclear vendors, and may not be achievable.

Figure 3.19
Demand for Construction and Outfitting Skills with 36-Month SSN Drumbeat

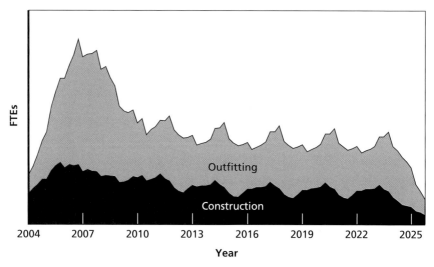

moved up in time by at least seven years; a 24-month drumbeat for the last five Astute-class boats suggests the MUFC should be accelerated by approximately 10 years.

Given the costs and risks involved, especially with the vendor base, we suggest that the SSN drumbeat be no longer than every 24 months and that the MUFC programme be accelerated to provide an in-service date of approximately 2026 if there is no follow-on to the Vanguard class. Such a strategy would provide a steady and continuous level of submarine production.

The above analysis is based on the assumption of eight SSNs in the force structure and an operational life of 25 years for the Astute-class and future submarines. Relaxing either of these assumptions could also help bridge the gap between the end of the eight-boat Astute build and the start of the MUFC class if there is no follow-on SSBN. Assuming a 24-month drumbeat is used, a force structure of 12 SSNs would permit continuous construction if submarines had operational lives of 25 years. If the force structure were held at eight submarines, reducing the operational life of the submarines to approximately 15 years would provide a steady construction drumbeat of one boat every 24 months.

Bridging the Gap in Submarine Production at Barrow

Downturns in submarine production of varying degrees could occur for different reasons at Barrow. Figure 3.18 depicts a fairly dramatic drop, but the situation in Figure 3.12 also amounts to a valley between production peaks.

If a downturn in submarine production should occur at Barrow, other shipbuilding work could be used to fill the valleys between peaks in submarine workload.[15] The UK shipbuilding industrial base is facing a growth in demand over the next decade from the CVF, MARS, and Type 45 programmes. There is some concern as to whether there is adequate capacity in the industrial base to meet these demands.[16] Barrow, with its long history of submarine and surface-ship construction, is a resource that may be needed to help the UK shipbuilding industrial base meet the planned build schedules of the various concurrent programmes.

Ships of the Type 45 class are currently under construction at the BAE Systems Clyde shipyards and the VT Shipbuilding shipyard. No decisions have yet been made on where the CVF and MARS ships will be built. The size of the CVF ships will require multiple shipyards to build major portions of the ship for assembly and completion of the construction at one location. Barrow may be able to contribute by building sections of the two CVF ships. As mentioned above, Barrow has recently completed the Wave-class auxiliary oilers and the Albion-class LPDs and has also constructed large commercial tankers in the past. These ships are similar to those in the MARS

---

[15] Even with a level demand in submarine work, other shipbuilding work could help reduce the overhead costs.

[16] Arena et al. (2005).

programme, and therefore Barrow could also participate in that programme. Of course, the MOD programmes will need to decide the best plans for the CVF and MARS programmes and for the overall MOD shipbuilding strategy. But increased demands in the face of reduced shipbuilding capacity may force some of this work to the Barrow shipyard.

As mentioned earlier in this chapter, there is the need for continuing nuclear-related work to maintain SQEP certifications. In the absence of suitable levels of nuclear submarine construction, experience could be gained through other nonnuclear naval work or by sending Barrow workers into other segments of the civilian nuclear industry or the nuclear vendors that support submarine construction. However, a return of these SQEP workers to nuclear submarine construction may require structured training plans for requalification.

In sum, non-submarine shipbuilding work at Barrow could help bridge a gap in submarine construction, thus providing workforce stability. Other shipbuilding work could also help to spread the overhead costs to various programmes, thereby reducing the costs of the submarines

Vitality of the Vendor Base[17]

With Barrow, the MOD must closely manage new production to ensure the vitality of the vendor base that provides equipment and materiel to submarine construction programmes. Many of these vendors are sole sources. We worked closely with BAE Systems Submarine Division to understand if any of the vendors previously supplying submarine production had quit, were experiencing problems, or might have difficulties in the future.

---

[17] Our analysis concentrates on the nonnuclear vendors that support submarine construction. A separate study by Rolls-Royce is examining the vitality of the nuclear vendors.

Identification of Departed or Potentially Problematic Suppliers

There are five suppliers that have provided materiel to submarine construction programmes but are no longer trading; they are shown in Table 3.5. Barrow has found replacements or potential replacements for these suppliers. In at least one case, the yard is considering manufacturing on its own the fabrications previously supplied by the vendor.

We identified 17 suppliers as potentially experiencing problems and rated these suppliers along two dimensions. First, we examined the long-term stability and viability of a supplier. With this measure, we hoped to understand whether or not the vendor would survive, given its total production base. Suppliers were rated low, medium, or high, with lower ratings indicating potential problems. The second measure addressed the supply risk: whether or not there were alternative providers if a supplier left the submarine industrial base. A high supply risk, denoting very few if any alternative sources for the product, indicated potential problems. We also sent surveys to the various suppliers asking about their overall business base and the portion of that base that was associated with submarine construction. The survey also asked the suppliers to assess their future in supporting submarine production and those actions that were necessary to help ensure their continued participation in the submarine industrial base. Five of the 17 companies responded to the survey.

Table 3.5
Suppliers No Longer Trading

Supplier	Product	Action
William Cook HiTech	NQ1 castings	Replaced by Sheffield Forgemasters
Motherwell Bridge	Special fabrications	Looking at in-house or alternative source
Thrust Engineering	Precision machining	Work transferred to Strand Engineering
Forward Industries	Hydraulics/actuators	Replaced by FCX Truflo
M&A Switchgear	Electrical equipment	Various alternatives identified

The 17 potentially problematic vendors are shown in Table 3.6. Only three companies were rated as having low stability. There are several alternative sources available for one of the products, so no real problems were identified for that product line. A second supplier was also rated as low in risk but had already been replaced. There are alternative suppliers for the third vendor, and Barrow is monitoring the situation to see whether an alternative supplier will be necessary.

Five companies were viewed as supply risks with few, if any, alternative sources of the equipment they provide. However, the stability of these companies is not a major concern at this time; they have sound corporate structures and other product lines to ensure their survivability.

Discussion

Our analysis of the nonnuclear vendors that support submarine construction suggests there are few current problems. This finding is con-

Table 3.6
Suppliers with Major Concerns or Issues

Supplier	Stability	Risk	Action/Comments
1	Medium	Medium	No major problems identified
2	Medium	Medium	Ongoing talks to restructure company
3	Low	Low	Financial problems
4	Medium	High	Company structure sound
5	Medium	Medium	No major problems as yet
6	Medium	Medium	No major problems as yet
7	Medium	Medium	No major problems as yet
8	Medium	Medium	No major problems as yet
9	Medium	High	No major problems as yet
10	Low	Low	Replaced
11	Medium	High	No major problems identified
12	Medium	Medium	Now have full order book
13	Medium	High	Company structure sound
14	Medium	Medium	No major problems as yet
15	Medium	High	Company structure sound Engineering short of work
16	Medium	Medium	Company structure sound
17	Low	Medium	Monitoring situation

sistent with our previous analyses of the vendor base that supports US submarine and aircraft carrier construction. It is also consistent with the observation that vendors were available to support the Astute programme after a several-year hiatus in submarine construction.

Typically, the vendor base that supports shipbuilding programmes raises concerns, since many vendors are considered sole sources and there is often variability in the demands for their products. However, the vendor base usually is found to be more robust than originally believed. Also, few nonnuclear vendors rely solely on shipbuilders as their only customers. In any case, shipbuilding orders represent only a small fraction of a company's total business. Furthermore, alternative sources are often available. As well, both the United Kingdom and the United States turn to foreign vendors if no domestic source is available.

Although the nonnuclear vendor base is currently stable, problems could occur in the future if there is another gap in submarine production or if the drumbeats are slow. The MOD may have to cope with higher equipment and material costs in the future if either of these situations occurs. There are solutions, though. For example, orders could be placed to cover a number of future submarines. This will keep vendors busy and may result in lower costs through higher order quantities. This action may not preclude problems in the future but will solve more immediate problems.

In the event of vendor shortfalls in the nonnuclear realm, the United Kingdom should consider working towards a common solution with the United States, which also faces potential vendor problems resulting from low production quantities. Perhaps one vendor, either a US firm or a UK firm, could be identified to support both the US and UK submarine construction programmes. As mentioned, this already happens to some degree today. Finally, the MOD could help by using the same vendors on both submarine and other shipbuilding programmes. For example, Strachan & Henshaw is actively pursuing work with the CVF programme.

Summary

Future nuclear submarine construction programmes must be carefully scheduled by the MOD. A long-term view is necessary to ensure submarines are produced in the most efficient manner. Although gaps in submarine production are possible, they are not preferred because they would result in loss of learning, lower productivity, and difficulty in maintaining certifications of key production personnel as suitably qualified and experienced. Workforce turbulence at Barrow can be reduced and labour-related costs lowered by carefully choosing programme end dates and drumbeats.

One key decision is whether there will be a future SSBN class once the Vanguard class reaches the end of its operational life. If there is not to be a follow-on to the Vanguard class, then it will be difficult or at least costly to sustain nuclear submarine production based solely on a limited number of attack submarines. To lessen the problem, either the size of the SSN fleet could be increased from eight to 12 boats or the operational life of a submarine could be reduced to approximately 15 years versus 25 years.

If plans are made for a follow-on SSBN submarine, then the next key decision is the operational life of the Vanguard-class submarines. There is some flexibility in this regard, since refuellings of the Vanguard class may allow operational lives ranging anywhere from 25 to 40 years. Production base issues should be considered in this decision. An operational life for the Vanguard class of between 31 and 34 years leaves sufficient time to build additional Astute-class submarines and to avoid a future gap before or an extensive overlap with the MUFC programme. If the objective is to allow enough of an overlap between the SSBN and MUFC programmes to permit a smooth transition, then the SSBN fleet size and drumbeat determine the programme start date.

The SSBN drumbeat must be chosen in combination with that for the remaining SSNs, as the relationship between them will affect the smoothness of the labour demand profile over the two programmes. The decision as to the number of attack submarines desired will also affect the SSN drumbeat (and thus the SSBN drumbeat).

For a fleet of eight Astute submarines (and assuming an SSBN requires 50 percent more labour hours than the Astute), an 18-month SSN drumbeat followed by a 36-month SSBN drumbeat will result in an almost uniform workforce demand at Barrow both for the total number of employees and for broad skill categories. Slower SSN drumbeats, particularly drumbeats slower than that for the SSBN, would result in a drop in workload demand at Barrow during SSN production followed by an increase in demand for SSBN production.

It might become necessary for budgetary or other reasons to delay for several years any further attack submarine production after the first three Astute-class boats. If so, a fleet of eight SSNs could not be sustained, although a fleet of six or seven SSNs could be. As mentioned, however, such a gap is not preferred.

If decisions are made that result in a decrease in submarine production, because SSN production is suspended or slowed down, then non-submarine shipbuilding work should be assigned to Barrow to help bridge that gap. Several new shipbuilding programmes are scheduled for the next decade, including CVF, MARS, and the Type 45 class of surface combatants. Barrow offers an option for meeting the increased demand for limited shipbuilding resources over the next decade.

There are currently no insurmountable problems with the non-nuclear vendor base that supports submarine construction. As with any set of vendors associated with the shipbuilding industrial base, some submarine-related vendors will exit the market. But new, substitute vendors are typically available. However, because production rates are expected to be low (even an 18-month drumbeat is not a high rate of production), costs for equipment and materiel could increase. This trend could be countered through economic order quantity buys (e.g., placing orders for multiple ship sets of equipment at one time), encouraging other MOD ship programmes to use submarine vendors, or using the vendors supporting other countries' submarine programmes.

CHAPTER FOUR

Summary Findings and Recommendations

The design and production of nuclear submarines requires special skills and unique leadership ability not found in other sectors of naval shipbuilding. Without adequate planning and a continuous level of work, these skills will atrophy and will be very difficult and costly to recreate. The small size of the Royal Navy's nuclear submarine fleet and the long time between designs for new classes of submarines requires the special attention of the MOD to ensure future submarines can be designed and built at reasonable costs.

Key Decisions to Sustain Nuclear Submarine Design Capability

Design and engineering resources at Barrow and at the major vendors are approaching precariously low levels as the design effort for the first three Astute-class submarines is concluding. The following decisions and actions are needed to sustain nuclear submarine design resources.

Decide If There Will Be a Next-Generation SSBN Class

A decision is needed soon on whether the Royal Navy will have a future underwater strategic nuclear deterrent to replace the current Vanguard-class SSBNs. If the decision is not to replace the Vanguard class, it will be very difficult and costly to sustain the UK submarine design and production base. Current defence plans call for eight nuclear attack submarines in the force structure. With an approximately 25-year operational life, new design programmes will be decades apart and new construction will be on the order of one submarine every 36 months. Those numbers will lead to very high costs for reconstituting nuclear submarine design resources and for very inefficient production rates. If there is no follow-on to the Vanguard class, then ways to reduce the potential damages to the nuclear submarine industrial base include increasing the number of SSNs in the force or reducing the operational life of an SSN.

Decide on the Operational Life of the Vanguard Class

If the decision is made to have a follow-on to the Vanguard class, the next decision would be when that follow-on should be ready.[1] There is some flexibility in this regard, as refuelling of Vanguard-class boats could extend their operational lives to anywhere between 25 and 40 years. Industrial base considerations could thus be a factor in the Vanguard replacement decision.

A 25-year operational life for the Vanguard class would require design efforts for the next-generation SSBN to begin immediately, which would create several problems. A near-term SSBN design programme would preclude the ability to enhance the design of the Astute class and would cause a gap of several years between the end of the follow-on SSBN design programme and the beginning of the MUFC design programme. It would also lead to competition for scarce design resources with other programmes such as the CVF, MARS, and the Type 45.

If the operational life of the Vanguard class were to be 40 years,[2] it would not be necessary (or desirable for other reasons) to start the design of the follow-on SSBN until well into the future. Such an effort might not start until a decade beyond the end of the current Astute contract. Even with additional design efforts for the remainder

---

[1] If there is no follow-on to the Vanguard, then the decision is when to start the next new SSN design programme. The findings are the same in either case.

[2] A 40-year operational life for the Vanguard class would require extensive, and expensive, updates of the combat systems on the boats.

of the Astute class, a sizeable gap would result. The delayed SSBN design effort would overlap with the design programme for the MUFC. There are currently insufficient design resources to support two such programmes simultaneously, and the same insufficiency is likely to prevail in the future, especially after a gap in design efforts of several years.

An operational life of slightly more than 30 years for the Vanguard class is preferable from the viewpoint of sustaining nuclear submarine design resources. This lifespan would leave several years for continuous development of the Astute class and allow that effort to blend well into the design programme for the MUFC.

Plan on Annual Investments to Sustain a Core of Design Resources
Historically, new nuclear submarine design programmes overlapped, allowing designers and engineers to shift from one programme to the next without committing to other projects. That trend was broken when there was a gap of several years between the end of the Vanguard programme and the beginning of the Astute programme. Even with careful planning of the type just described, small fleet sizes and long operational lives will lead to future gaps of at least a few years' duration between new nuclear submarine design programmes. Efforts are needed to sustain design resources between such gaps.

The MOD should plan on providing funding to sustain a core of approximately 200 designers, engineers, and draughtsmen. This would require annual funding of approximately £15 million. The funding should support long-range conceptual design studies to evaluate new technologies and to encourage innovative thinking. It should also support in the near-term continuous development efforts on the Astute class. These efforts should be oriented to incorporating new technologies and to reducing the production and ownership costs of the Astute class specifically and nuclear submarines in general. Finally, modest funding should support continuing designs for UUVs.

Begin Talks with the United States on Collaborative Design Programmes

Reduced force sizes and budget constraints cause difficulties in sustaining nuclear submarine design resources in both the United Kingdom and the United States. Collaboration between the two countries can lead to numerous benefits. With some level of reliance on each other, both countries can reduce the burden of maintaining high levels of resources between new design programmes and can draw on the other country's design resources when a new programme begins. Collaboration can also help the interoperability of the countries' submarine forces and can lead to a sharing of skills and expertise that one country or the other now lacks.

Any collaborative effort in the design of nuclear submarines must overcome the hurdles of intellectual property rights and the sharing of classified and other sensitive information. Talks should start immediately to set the boundaries for a collaborative effort and to facilitate the sharing of necessary information. These talks must begin at the government level but must also involve the private companies that design and produce nuclear submarines (i.e., BAE Systems Submarine Division, Electric Boat, and Northrop Grumman Newport News).

Collaborative efforts could start small, possibly a few dozen designers and engineers from each country working together. Because both countries are potentially facing new SSBN designs, collaborative efforts could be directed at conceptual studies for follow-on SSBN classes. Both countries are also producing the first of a new class of attack submarines. Collaborative efforts could be directed at spiral development activities for those new classes. Finally, both countries are expending resources for the design and prototyping of new UUVs, which could also benefit from collaboration. Such efforts would be primarily intended to test the collaborative concept and solve impediments to implementation.

From a small start, however, collaboration could eventually grow to approximately 200 or more designers and engineers from one country working on a new nuclear submarine design programme in the other country. For that to prove helpful in sustaining a design core, the new programmes of the two countries would have to be off-set at some point. Also, a common software design tool may be needed to facilitate the collaborative efforts.

Encourage Collaboration Between the Various Design Organisations
The MOD should also encourage collaboration among the various UK organisations that provide design resources for new submarine programmes and those that provide design resources to support in-service submarines. BAE Systems Submarine Division, DML, and BES all have a number of highly skilled nuclear submarine designers and engineers. These organisations should start working together on both new design programmes and the support of fleet submarines. There is insufficient demand across nuclear submarine design, pro-duction, and support to continually sustain large numbers of design professionals at each organisation. DML and BES designers and engineers should be part of any new submarine design programme, bringing their general knowledge of submarine design plus their spe-cific knowledge of the support of in-service submarines. Likewise, Barrow designers and engineers should work with DML and BES on the in-service support of the Astute class, especially if they maintain design authority once the class enters the active fleet.

Although the focus of our study was on sustaining nuclear sub-marine design and production resources, we note the potential future problem of sustaining general shipbuilding design skills in the United Kingdom. Several new design programmes have started or will soon start. These include the Type 45, CVF, and MARS as well as the Astute class. When the design efforts for these programmes are com-pleted, there is potentially a decade or more before new shipbuilding design programmes will be needed. Just as collaboration with the United States would help sustain nuclear submarine design resources in both countries, collaboration between the various organisations involved in new ship designs would help sustain those resources. Fur-ther study is warranted to determine how such collaboration could be managed to produce new ship designs.

Key Decisions to Sustain Nuclear Submarine Production Capability

Like design, the production of nuclear submarines requires unique skills and training. Submarine construction gaps differ from those in the construction of surface ships in some basic ways. Gaps in the building of nuclear submarines, even if those gaps are filled with non-submarine construction work, can lead to loss of learning in specialised skills and increased production costs. Also, SQEP must maintain their qualifications through demonstrating their ability to adequately perform production tasks on nuclear submarines. Finally, many of the vendors that support nuclear submarine construction, especially those that support the nuclear steam-raising plant, are sole sources whose vitality may be threatened if there are gaps in submarine production.

The MOD must take a long-term view of nuclear submarine production schedules to ensure the adequacy of the industrial base and to obtain the best value for money. It must look beyond the next submarine, or the next submarine contract, and plan for the smooth integration of future programmes. This requires understanding, within and across programmes, the interactions among decisions regarding the timing of construction start or end, the number of boats required, and the pace of new starts (the drumbeat). The framework we present here for relating these factors should assist the MOD in making those types of decisions.

Decide on the Operational Life of the Vanguard Class

Looking to the future, a decision is needed on when the next new submarine programme will start. This is similar to, and related to, the design issues described above. Assuming the next new nuclear submarine programme is for a follow-on SSBN class, the immediate decision is again the planned operational life of the Vanguard class. For reasons that parallel those from a design perspective, the operational life of the Vanguard class should, from a production perspective, be between 31 and 34 years. This will allow a smooth transition from the production of the remaining boats in the Astute class to the production of the follow-on SSBN class and then to the construction of the MUFC class submarines. The efficient drumbeat for the new SSBN class is related to decisions on the remaining Astute-class production. In general, the follow-on SSBN-class production drumbeat should be slower than the Astute-class drumbeat (since the follow-on SSBN will require more production hours than an Astute-class boat).

Decide on Follow-On Astute-Class Production
It has been announced that the future Royal Navy force size will be eight submarines. This requires five more Astute-class boats to be built to replace retiring Swiftsure- and Trafalgar-class submarines as they retire. Decisions must be made on when to start the remaining Astute-class construction and what drumbeat to use in building the boats.

For the efficient use of the production resources, the fourth Astute-class boat should start construction in the next year or two. Given the goal of an eight-boat fleet, the production drumbeat for the remaining five boats should be between 18 and 24 months. The least variable long-term demand for production workers would be provided by an 18-month drumbeat, followed by an SSBN drumbeat of 30 or 36 months. A slower production drumbeat for the remaining boats in the Astute class would result in more of a drop in workforce levels and then a ramp-up for the follow-on SSBN production; this would be particularly true if the Astute drumbeat is slower than that for the SSBN.

Even with the best of efforts to schedule submarine production to minimise production gaps, some lulls in production may be unavoidable, or periods of lower activity may fall between periods of greater activity. That is particularly likely if a 24-month or slower drumbeat is used for the remaining boats in the Astute class. Non-submarine shipbuilding work might help fill such demand valleys at Barrow. Not only would such work help meet an expected mid-term national demand peak for surface-ship production resources, it should also lead to a reduction in the cost of producing nuclear submarines through spreading overhead and reducing the workforce turbulence costs. Programmes that might assign work to Barrow include the CVF and MARS.

Take Actions to Support Nonnuclear Vendors

Currently, there are no significant problems with the survivability of the nonnuclear vendors that support nuclear submarine construction. However, even the quicker drumbeats recommended here are slow enough to possibly place stresses on the vendor base and may lead to higher equipment and materiel costs. The MOD should take actions to bolster the submarine vendors and to control rising costs. These actions include placing orders for multiple ship sets (for at least the last five Astute-class submarines) where practical, encouraging other MOD shipbuilding programmes to use the submarine vendors, and working with the United States to identify common vendors to support both countries' submarine programmes.

A Brief History of UK Submarine Production

The Beginning: 1886 Through World War II

The history of UK submarine production and the history of production of submarines at Barrow are nearly synonymous. The shipyard at Barrow was originally established in 1872 as the Barrow Ship Building Company. The first Barrow-built submarine, the Nordenfelt, was produced in 1886. 'This steam-driven boat was … an improved version of an earlier submarine built in Stockholm in 1882.'[1] There was one more Nordenfelt built in 1887 and then a hiatus in production until 1901. This production began after the yard was bought by Vickers in 1897, and was the result of an order from the Holland Torpedo Boat Company of America, currently known as the Electric Boat Corporation. Working drawings were supplied to Vickers by Electric Boat, and construction of what is now called the Holland class began. However, during the construction programme, the relationship between Electric Boat and Barrow was dissolved, the postulated reason being communication difficulties. Adjustments were made to the submarine design and the result was the Holland class of five submarines in 1903. Some would consider these to be the first submarines produced, and they were the first produced in the yard while under the ownership of Vickers, Sons & Maxim.

The remainder of submarine production up through the end of World War II is characterised by rapid production, spiral develop-

---

[1] The Barrow-in-Furness Submarines Association's Web site, www.submariners.co.uk.

ments,[2] and relatively short in-service lives. Very few were in the fleet for more than 15 years.

There were more than 479 submarines produced in the United Kingdom from 1901 to the end of World War II, most of which were built at Barrow[3] (see Table A.1). During peak production in 1942, an average of more than two boats per month were produced. This was a period in which two World Wars drove production requirements, and the cost of development, though not inconsequential, was considered well worth the product. In many cases, the completion of the current class was recently under way when errors or weaknesses were discovered and the decision was made to begin the next class. Many of the submarines produced during this time never made it into service because of technical problems; others did make it into service but were quickly retired because they had technical problems or did not provide sufficient capability. However, each submarine helped the country to advance to the next class of submarines. The first 45 years of production resulted in the establishment of a strong foundation for the production of submarines.

The United Kingdom continued building submarines after the war, and by the 1950s had a stable diesel-electric submarine shipbuilding capability in several shipyards, including Chatham, Cammell Laird, Vickers-Armstrong Limited, and Scott's Shipbuilding and Engineering Company Limited, Greenock. After World War II, the United Kingdom also briefly pursued the use of hydrogen peroxide as a propulsion power source with the submarines Excalibur and Explorer, but most of their submarines continued with the diesel-powered approach.

---

[2] According to Major Ross McNutt, Acquisition Management Policy Division, Office of the Secretary of the Air Force, *spiral development* is a 'method or process for developing a defined set of capabilities, providing opportunity for interaction between the user, tester, and developer communities to refine the requirements, provide continuous feedback and provide the best possible capability process of continually updating requirements based on feedback from the client'.

[3] Data received from Barrow-in-Furness Submariners Association's Web site, www.submariners.co.uk.

Table A.1
History of Submarine Production at Barrow:
1886 to Pre–World War II

Class Name	Number in Class	First-of-Class In-Service Date	Last Boat Out-of-Service Date
Nordenfelt	2	1886	1890
Holland	5	1901	1913
A	13	1902	1908
B	11	1903	1906
C	38	1905	1910
D	8	1907	1919
E	57	1911	1924
V	4	1912	1919
G	14	1914	1921
K	17	1915	1931
K26	1	1918	1931
M	3	1916	1932
L	27	1916	1945
H21	13	1918	1945
R	10	1917	1934
Patrol Submarines	19	1925	1946
River	3	1929	1945
Minelaying	6	1930	1946
S	62	1935	1970
T	53	1935	1970
U	49	1936	1950
V	22	1941	1958
Midget	20	1939	1952

The Porpoise and Oberon classes (Tables A.2 and A.3) followed the same cycle of design improvement as the World War II–class diesel submarines. For example, the Oberon class was designed for greater battery capacity and improved submerged performance. The emphasis was placed on reducing noise radiating from the submarine, in parallel with a shift in submarine target detection methods to passive sonar listening. As a result, the Oberon boats were the quietest UK submarines to date.[4] The Oberon was first laid down in November 1957, with the last in class completed in 1967.

---

[4] Ray Burcher and Louis Rydill, *Concepts in Submarine Design*, Cambridge, UK: Cambridge University Press, 1995, p. 19.

Table A.2
Porpoise-Class Submarines (Diesel)

Name	Builder	Laid Down	Completed
Porpoise (S01)	Vickers-Armstrong	June 1954	April 1958
Rorqual (S02)	Vickers-Armstrong	January 1955	October 1958
Grampus (S04)	Cammell Laird	April 1955	December 1958
Cachalot (S06)	Scott's	August 1955	September 1959
Narwhal (S03)	Vickers-Armstrong	March 1956	May 1959
Finwhale (S05)	Cammell Laird	September 1956	August 1960
Walrus (S08)	Scott's	February 1958	February 1961
Sealion (S07)	Cammell Laird	June 1958	July 1961

Table A.3
Oberon-Class Submarines (Diesel)

Name	Builder	Laid Down	Completed
Oberon (S09)	H.M. Dockyard, Chatham	November 1957	February 1961
Odin (S10)	Cammell Laird	April 1959	May 1962
Onslaught (S14)	H.M. Dockyard, Chatham	April 1959	August 1962
Orpheus (S11)	Vickers-Armstrong	April 1959	November 1960
Otter (S15)	Scott's	January 1960	August 1962
Olympus (S12)	Vickers-Armstrong	March 1960	July 1962
Oracle (S16)	Cammell Laird	April 1960	February 1963
Ocelot (S17)	H.M. Dockyard, Chatham	November 1960	January 1964
Opossum (S19)	Cammell Laird	December 1961	June 1964
Otus (S18)	Scott's	May 1961	October 1963
Osiris (S13)	Vickers-Armstrong	January 1962	January 1964
Opportune (S20)	Scott's	October 1962	December 1964
Onyx (S21)	Cammell Laird	November 1964	November 1967

The Advent of the Nuclear Submarine

Submarine propulsion technology underwent a revolutionary change in the United States when the first nuclear submarine, the USS *Nautilus,* went to sea in 1954. By the mid-1950s, the United Kingdom had established its own nuclear propulsion research and development programme at Harwell. Vickers Limited shipyard at Barrow was to be a major partner.[5] The United Kingdom selected pressurised water reactor technology as the system of choice for its nuclear propulsion system and set a target date of 1961 for the launch of the HMS *Dreadnought*. For this boat, the Ministry of Defence (MOD) led the nuclear submarine concept development efforts. Vickers Nuclear Engineering Limited provided constructor interface and insight, and the naval section at Harwell provided overall programme coordination.

As development of an independently designed UK nuclear propulsion plant proceeded, the United Kingdom began discussions with the United States about the possibility of transferring US-developed nuclear technology to the United Kingdom. The result was that the United States agreed to provide a complete and proven Skipjack-class reactor plant, with supporting documentation, to the United Kingdom for installation in the Dreadnought. The United Kingdom decided that Rolls-Royce and Associates would be the single point of contact to the Westinghouse Corporation, the US provider of nuclear equipment, and the transfer would be managed through that linkage. Vickers-Armstrong built Dreadnought at Barrow. Its keel was laid down in 1955. The submarine was launched in 1960 and commissioned in April 1963.

After the completion of Dreadnought, the United Kingdom embarked on the first entirely British-designed and -developed nuclear submarine, the Valiant class (Table A.4). The first Valiant was laid down in January 1962, with the last in class completed in November 1971. In early 1963, the United Kingdom announced that it planned to order four 7,000-ton Resolution-class (Table A.5) nuclear-powered ballistic missile submarines to deploy a strategic nuclear deterrent, with the aim of putting them on patrol beginning in 1968. Vickers-Armstrong was hired to build two submarines and

---

[5] Y-ARD, a design support contractor, was a major partner as well.

Table A.4
Valiant-Class Submarines

Name	Builder	Laid Down	Completed
Valiant (S102)	VSEL, Barrow	January 1962	July 1966
Warspite (S103)	VSEL, Barrow	December 1963	April 1967
Churchill (S46)	VSEL, Barrow	June 1967	July 1970
Conqueror (S48)	Cammell Laird	December 1967	November 1971

Table A.5
Resolution-Class Submarines

Name	Builder	Laid Down	Completed
Resolution (S22)	VSEL, Barrow	February 1964	October 1967
Repulse (S23)	VSEL, Barrow	March 1965	September 1968
Renown (S26)	Cammell Laird	June 1964	November 1968
Revenge (S27)	Cammell Laird	May 1965	December 1969

provide lead yard service to the builder of the other two. The first in class was laid down in February 1964, with the last completed in December 1969.

The four Resolution-class submarines were followed by Swiftsure (Table A.6) attack submarines, and then by the Trafalgar class (Table A.7). Each of these represented an improvement in technology and capability. The first Swiftsure-class submarine was laid down in June 1969, with the last being completed in March 1981. The Trafalgar class followed immediately, with the first being laid down in April 1979 and the last completed in October 1991.

In parallel with the construction of its nuclear submarine flotilla, the United Kingdom introduced the Upholder class of four conventional (diesel)–powered submarines. These were built in the mid-1980s through the early 1990s. The Upholder class (Table A.8) included the last submarines built in the United Kingdom outside the Barrow shipyard. The first was laid down in November 1983, with the last completed in June 1993.[6]

---

[6] The four Upholder-class submarines have been recently sold to Canada.

Table A.6
Swiftsure-Class Submarines

Name	Builder	Laid Down	Completed
Swiftsure (S126)	VSEL, Barrow	June 1969	April 1973
Sovereign (S108)	VSEL, Barrow	September 1970	July 1974
Superb (S109)	VSEL, Barrow	March 1972	November 1976
Sceptre (S104)	VSEL, Barrow	February 1974	February 1978
Spartan (S105)	VSEL, Barrow	April 1976	September 1979
Splendid (S106)	VSEL, Barrow	November 1977	March 1981

Table A.7
Trafalgar-Class Submarines

Name	Builder	Laid Down	Completed
Trafalgar (S107)	VSEL, Barrow	April 1979	May 1983
Turbulent (S87)	VSEL, Barrow	May 1980	April 1984
Tireless (S88)	VSEL, Barrow	June 1981	October 1985
Torbay (S90)	VSEL, Barrow	December 1982	February 1987
Trenchant (S91)	VSEL, Barrow	October 1985	January 1989
Talent (S92)	VSEL, Barrow	May 1986	May 1990
Triumph (S93)	VSEL, Barrow	February 1987	October 1991

Table A.8
Upholder-Class Submarines (Diesel)

Name	Builder	Laid Down	Completed
Upholder (S40)	VSEL, Barrow	November 1983	June 1990
Unseen (S41)	Cammell Laird	January 1986	June 1991
Ursula (S42)	Cammell Laird	August 1987	May 1992
Unicorn (S43)	Cammell Laird	February 1989	June 1993

In July 1980, the United Kingdom decided to modernise its strategic nuclear deterrent through the purchase of the United States' Trident (C4) missile system. This acquisition was followed by an announcement in 1982 that the government had opted for the upgraded Trident II system to be deployed at sea on a force of four submarines by the middle of the following decade. The Vanguard class (Table A.9) was developed to deploy this weapon system. The first was laid down in September 1992, with the last boat in the class completed in November 1999.

Currently, the UK submarine fleet is composed of the four Vanguard-class SSBNs, the seven Trafalgar-class SSNs, and four Swiftsure-class boats.

In the late 1980s, the MOD began conducting a series of studies for a follow-on attack submarine, the SSN20, which would eventually replace the Swiftsure class and then the Trafalgar class. The intention was to follow the Trafalgar class with an improved class of SSNs. However, while SSN20 was in its initial stages of development, the fall of the Berlin wall and subsequent disintegration of the Soviet Union altered national perspective to the point where the UK Treasury questioned the high cost of the improved capability. This led to a new approach for a follow-on submarine, which was characterised as the 'Batch 2 Trafalgar class' that was eventually labelled the Astute. Contracts for the studies phase of this renamed submarine were placed in 1992 and 1993.

Further contracts for risk reduction were placed with two competing contractors, VSEL and GEC Marconi. Invitations to tender were issued in 1994, with bids received in 1995. After a further series of studies and a competition, the contract was issued to GEC Marconi in 1997. GEC Marconi then bought VSEL and was in turn acquired by BAE Systems, the current prime contractor for the submarine. The keel was laid down in January 2001, with launch expected in 2009.

Table A.9
Vanguard-Class Trident Missile Submarines

Name	Builder	Laid Down	Completed
Vanguard (S28)	VSEL, Barrow	September 1986	August 1993
Victorious (S29)	VSEL, Barrow	December 1987	January 1995
Vigilant (S30)	VSEL, Barrow	February 1993	November 1996
Vengeance (S31)	VSEL, Barrow	February 1993	November 1999

During production of the Swiftsure, Trafalgar, and Vanguard classes, a new submarine was being delivered to the Royal Navy every one or two years (see Figure A.1). That pattern will be broken with the Astute class. Under current plans, the first Astute-class submarine will be delivered eight years after the last of the Vanguard class is delivered. The gap in submarine production at Barrow is one reason cited for the cost and schedule problems facing the Astute class.

UK Shipbuilding Nationalisation and Ownership of Barrow

A brief history of UK submarine production would not be complete without commenting on the turbulence that has been characteristic of ownership of Barrow. Nationalisation of the shipyards and a series of mergers and acquisitions resulted in continuously changing owner-ship of Barrow. Vickers Shipbuilding owned Barrow until 1995.

Figure A.1
Time Between Deliveries of Nuclear Submarines (1975 to Present)

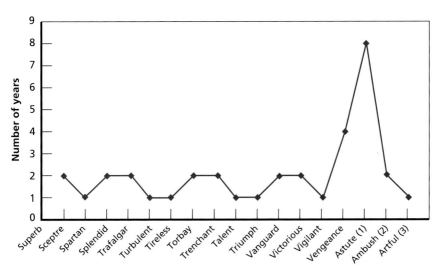

RAND *MG326/1-A.1*

During the period in which Vickers owned the yard, the exact name of the company of ownership changed at least eight times, all in association with Vickers.[7] When shipbuilding was privatised in 1986, the status of the Vickers Company changed three times, first as Vickers Shipbuilding & Engineering Limited (a subsidiary of British Shipbuilders), second as VSEL Consortium PLC, and in August 1986 it became VSEL PLC. In 1995, GEC Marconi purchased Vickers Shipbuilding & Engineering Limited.

Several events occurred in the short time that GEC Marconi owned the yard. Shortly before the purchase of the yard, VSEL had launched two Vanguard-class submarines, one in 1992 and one in 1993. The Vigilant submarine was launched in 1995, the year of the GEC Marconi purchase, and the last of the Vanguard class, the Vengeance, was launched in 1998, three years after the purchase of the yard. The next year, in 1999, BAE Systems bought the Barrow shipyard. When the yard was purchased, the responsibility for producing the Astute class was transferred to BAE Systems through the necessary contracts.

---

[7] Barrow Island Web site, www.barrow-island.com.

The Nuclear Submarine Design Process

The historical design process consisted of four phases: concept design, preliminary design, contract design, and detailed design.[1]

Concept Design

In this phase, concepts are explored against a backdrop of a continuing evaluation of future missions, future threats, and future technologies. The dialogues between the designer and the operator/requirements planner and between the designer and the technologist are crucial. Systems designers with a broad view of technology, current and future operational planning, and operational experience with current designs are necessary to provide leadership during the conceptual design phase. As concepts are explored and defined, trade-offs are made with military effectiveness, affordability, and producibility as the principal driving system criteria. The output of the concept design phase is the definition of a preferred design concept by a set of 'single sheet' characteristics, stipulating submarine missions, principal operating and performance characteristics and dimensions, military

---

[1] The MOD acquisition cycle consists of the Concept, Assessment, Demonstration, Manufacture, In-Service, and Disposal (CADMID) phases. Concept design is accomplished during the Concept phase. Preliminary and contract design occur during the Assessment phase. Detailed design is part of the Demonstration and Manufacture phases. For a description of the CADMID cycle, see UK Ministry of Defence, *The Smart Acquisition Handbook*, Edition 5, Director General Smart Acquisition Secretariat, January 2004.

payload, and design affordability and producibility goals. An estimate of the cost of construction is also developed.

Preliminary Design

During this phase, the preferred concept is matured. Subsystem configurations and alternatives are examined and analysed for military effectiveness, affordability, and producibility. Trade-off decisions are made and analysis and testing are performed in areas such as structures, hydrodynamics, silencing, combat system performance, and arrangements. The output of preliminary design is a set of top-level requirements explicitly describing the refinements achieved during this phase. The performance requirements are established in sufficient detail. The characteristics and ship dimensions are also spelled out in far more detail than the 'single sheet' characteristics of the conceptual design phase. Also, a more refined budget estimate is provided as input to the budget cycle.

Contract Design

This phase consists of the transformation of the top-level requirements into contracts for the detailed design and construction of the submarine. All subsystems are defined; all analysis and testing is completed; the projected cost of the detailed design and construction of the submarine is established; and a set of ship specifications and contract drawings are prepared. This enables an 'invitation to tender' document to be issued so that the shipbuilder can respond with proposals that form the basis for the negotiation of the price and terms and conditions of the final contract.

Detailed Design

This phase is normally performed by the shipbuilder, since it transforms the contract drawings and ship specifications into the documents necessary to construct, outfit, and test the submarine. Typical products consist of working drawings, work orders, test memoranda, shipyard procedures, erection sequences, and the like. Construction of the submarine often starts long before the detailed design is complete. Too much of an overlap of detailed design and construction can limit the efficient construction of the submarine.

Modern Changes to the Design Process

During the traditional approach described above, each of the phases of design was performed independently of each other with often a gap between the phases as decisions were being made. The modern integrated product and process design (IPPD) process blends the four traditional phases together into an almost seamless process.[2] Now, a system definition phase occurs followed by an integrated design/construction planning development phase. This change has resulted in complete designs much earlier than under the traditional design process.

An important change results from the IPPD process. Construction does not start until the detailed design drawings are largely complete. This allows more efficient production and reduces the number of changes made during the construction process. Electric Boat also champions the design/build process in which people knowledgeable of the construction process are incorporated into the design process to help increase the producibility and cost effectiveness of the design.

---

[2] For an excellent description of the application of IPPD to nuclear submarine design and construction, see General Dynamics Electric Boat, *The VIRGINIA Class Submarine Program: A Case Study*, Groton, Conn., USA, February 2002.

Bibliography

Arena, Mark V., Hans Pung, Cynthia R. Cook, Jefferson P. Marquis, Jessie Riposo, and Gordon T. Lee, *The United Kingdom's Naval Shipbuilding Industrial Base: The Next Fifteen Years*, Santa Monica, Calif., USA: RAND Corporation, MG-294-MOD, 2005.

Arena, Mark V., John F. Schank, and Megan Abbott, *The Shipbuilding & Force Structure Analysis Tool: A User's Guide*, Santa Monica, Calif., USA: RAND Corporation, MR-1743-NAVY, 2004.

Bennett, John T., 'Navy Eyes Enhanced UUV for Future Littoral Combat Ship Deployment', Inside Washington Publishers, 8 April 2004.

Birkler, John, Joseph P. Large, Giles K. Smith, and Fred Timson, *Reconstituting a Production Capability: Past Experience, Restart Criteria, and Suggested Policies*, Santa Monica, Calif., USA: RAND Corporation, MR-273-ACQ, 1993.

Birkler, John, John Schank, Giles K. Smith, Fred Timson, James Chiesa, Marc D. Goldberg, Michael Mattock, and Malcolm MacKinnon, *The U.S. Submarine Production Base: An Analysis of Cost, Schedule, and Risk for Selected Force Structures*, Santa Monica, Calif., USA: RAND Corporation, MR-456-OSD, 1994.

Birkler, John, Michael Mattock, John Schank, Fred Timson, James Chiesa, Bruce Woodyard, Malcolm MacKinnon, and Denis Rushworth, *The U.S. Aircraft Carrier Industrial Base: Force Structure, Cost, Schedule, and Technology Issues for CVN 77*, Santa Monica, Calif., USA: RAND Corporation, MR-948-NAVY/OSD, 1998.

Birkler, John, Denis Rushworth, James Chiesa, Hans Pung, Mark V. Arena, and John F. Schank, *Differences Between Military and Commercial Ship-*

building: Implications for the United Kingdom's Ministry of Defence, Santa Monica, Calif.: RAND Corporation, MG-236-MOD, 2005.

Birkler, John, John F. Schank, Mark Arena, Giles K. Smith, and Gordon Lee, *The Royal Navy's New-Generation Type 45 Destroyer: Acquisition Options and Implications*, Santa Monica, Calif., USA: RAND Corporation, MR-1486-MOD, 2002.

Burcher, Ray, and Louis Rydill, *Concepts in Submarine Design*, Cambridge, UK: Cambridge University Press, 1995.

Butler, Nicola, and Mark Bromley, *Secrecy and Dependency: The UK Trident System in the 21st Century*, BASIC, Number 2001.3, November 2001. Online at www.basicint.org/pubs/Research/2001UKtrident1.htm (accessed March 2005).

Cook, Cynthia R., John Schank, Robert Murphy, James Chiesa, John Birkler, and Hans Pung, *The United Kingdom's Nuclear Submarine Industrial Base, Volume 2: MOD Roles and Required Technical Resources*, RAND Corporation, MG-326/2-MOD, forthcoming.

Electric Boat Corporation, *Shipbuilding Industrial Base Study*, private/ proprietary information, 29 August 1997.

_____, *Shipbuilding Engineering Resources*, AD-20481, 30 March 1999.

'Executive Overview', *Jane's Underwater Technology*, 12 January 2004.

Funnell, Clifford, 'Update History: Analysis', *Jane's Underwater Technology*, 12 January 2004.

Furness Enterprise, *UK Submarine Industrial Base: Employment Issues Associated with Nuclear Steam Raising Plant Installation in New Submarines*, 8 July 2004.

General Dynamics Electric Boat, *The VIRGINIA Class Submarine Program: A Case Study*, Groton, Conn., USA, February 2002.

Hartley, Keith, *The UK Submarine Industrial Base: An Economics Perspective*, York, England: Centre for Defence Economics, University of York, no date.

_____, *The Economics of UK Procurement Policy*, York, England: Centre for Defence Economics, University of York, October 2002.

Lacroix, F. W., Robert W. Button, John R. Wise, and Stuart E. Johnson, *A Concept of Operations for a New Deep-Diving Submarine*, Santa Monica, Calif., USA: RAND Corporation, MR-1395-NAVY, 2001.

Ma, Jason, 'ONR Developing Two Unmanned Sea Surface Vehicle Prototypes', *Inside the Navy*, 26 July 2004.

Milk, Keith, *The Regeneration of UK Defence Forces: A Modeling Approach*, HVR Consulting Services Limited, PV/90/004, 9 September 2002.

_____, *The Future Regeneration of the Submarine Industrial Base: Phase 1 Report*, HVR Consulting Services Limited, AW/05/003 Issue 1.0, 29 April 2003.

_____, *The Future Regeneration of the Submarine Industrial Base: Phase 2 Report*, HVR Consulting Services Limited, AW/05/006 Issue 1.0, 17 October 2003.

'Mine Warfare: Underwater Vehicles, United Kingdom', *Jane's Underwater Systems*, 10 June 2004.

Newport News Shipbuilding, *Reconstitution of the Submarine Work Force, Revision B*, company confidential, 1 June 1993.

Pugh, Philip, and D. Faddy, *Reconstitution of U.K. Defence Forces*, HVR Consulting Services Limited, MV/07is/010, 16 January 2003.

Raman, Raj, Robert Murphy, Laurence Smallman, John Schank, John Birkler, and James Chiesa, *The United Kingdom's Nuclear Submarine Industrial Base, Volume 3: Options for Initial Fuelling*, Santa Monica, Calif., USA: RAND Corporation, MG-326/3-MOD, 2005.

Schank, John, John Birkler, Eiichi Kamiya, Edward Keating, Michael Mattock, Malcolm MacKinnon, and Denis Rushworth, *CVX Propulsion System Decision: Industrial Base Implications of Nuclear and Non-Nuclear Options*, Santa Monica, Calif., USA: RAND Corporation, DB-272-NAVY, 1999.

Scott, Richard, 'Recovery Plan Shapes Up for Astute Submarines', *Jane's Defence Weekly*, 19 November 2003.

UK Ministry of Defence, *The Smart Acquisition Handbook*, Edition 5, Director General Smart Acquisition Secretariat, January 2004.

_____, 'Delivering Security in a Changing World: Future Capabilities', white paper presented to Parliament by the Secretary of State for Defence, July 2004.

US Department of the Navy, The Navy Unmanned Undersea Vehicle (UUV) Master Plan, 20 April 2000.